Trusting the Subject?

The Use of Introspective Evidence
in Cognitive Science

Volume 1

Edited by

Anthony Jack

&

Andreas Roepstorff

ia

IMPRINT ACADEMIC

OF RELATED INTEREST FROM IMPRINT ACADEMIC
Full details on: http://www.imprint-academic.com

Series Editor: Professor J.A. Goguen
Department of Computer Science and Engineering, UCSD

Francisco J. Varela and Jonathan Shear, ed.
The View From Within:
First-Person Approaches to the Study of Consciousness

Evan Thompson, ed.
Between Ourselves:
Second-Person Issues In the Study of Consciousness

Michel Ferrari, ed.
The Varieties of Religious Experience: Centernay Essays

Alva Noë, ed.
Is the Visual World a Grand Illusion?

Published in the UK by Imprint Academic
PO Box 200, Exeter EX5 5YX, UK

Published in the USA by Imprint Academic
Philosophy Documentation Center,
PO Box 7147, Charlottesville, VA 22906-7147, USA

World Copyright © Imprint Academic, 2003
No part of any contribution may be reproduced in any form without permission,
except for the quotation of brief passages in criticism and discussion.

ISBN 0 907845 56 8 (paperback)

ISSN 1355 8250 (*Journal of Consciousness Studies*, **10**, number 9–10, 2003)

British Library Cataloguing in Publication Data
A catalogue record for this book is available from the British Library
Library of Congress Card Number: 2003102065

Cover Art: stevek@conceptstudio.co.uk

Contents

Anthony Jack and Andreas Roepstorff,
 Editorial Introduction ... v

K. Anders Ericsson,
 Valid and Non-Reactive Verbalization of Thought During Performance of Tasks: Towards a Solution to the Central Problems of Introspection as a Source of Scientific Data ... 1

Daniel C. Dennett,
 Who's On First? Heterophenomenology Explained ... 19

Antoine Lutz and Evan Thompson,
 Neurophenomenology: Integrating Subjective Experience and Brain Dynamics in the Neuroscience of Consciousness ... 31

Dan Zahavi and Josef Parnas,
 Conceptual Problems in Infantile Autism Research: Why Cognitive Science Needs Phenomenology ... 53

Patrick Haggard and Helen Johnson,
 Experiences of Voluntary Action ... 72

Shaun Gallagher,
 Phenomenology and Experimental Design: Toward a Phenomenologically Enlightened Experimental Science ... 85

Bernard J. Baars,
 How Brain Reveals Mind: Neural Studies Support the Fundamental Role of Conscious Experience ... 100

David Leopold Alexander Maier and Nikos K. Logothetis,
 Measuring Subjective Visual Perception in the Nonhuman Primate ... 115

Timothy D. Wilson,
 Knowing When to Ask: Introspection and the Adaptive Unconscious ... 131

Gualtieri Piccinini,
 Data from Introspective Reports: Upgrading from Common Sense to Science ... 141

Richard E. Cytowic,
 The Clinician's Paradox: Believing Those You Must Not Trust ... 157

Anthony J. Marcel,
 Introspective Report: Trust, Self Knowledge and Science ... 167

Contributors

Bernard J. Baars (3616 Chestnut St. Apt. 3, Lafayette, CA 94549, USA. baars@nsi.edu) is an Affiliated Research Fellow at the Neurosciences Institute in San Diego, engaged in collaborative writing projects with various researchers, such as *Essential Sources in the Scientific Study of Consciousness* (2003), co-edited with William Banks and the late James Newman. His integrative theory of consciousness, global workspace theory, was detailed in *A Cognitive Theory of Consciousness* (1988), with a briefer popular account in *In the Theater of Consciousness* (1997). Recent articles are available on www.nsi.edu/users/baars.

Richard E. Cytowic (4720 Blagden Terrace NW, Washington DC 20011-3720, USA. Richard@Cytowic.net) is an MD credited with returning synaesthesia to the scientific mainstream. His textbooks and popular writings include *Synesthesia: A Union of the Senses*, *The Neurological Side of Neuropsychology*, and *The Man Who Tasted Shapes*. His degrees come from Duke University and Wake Forest's Bowman Gray School of Medicine. Founder of Capitol Neurology in Washington, DC, he is now retired from clinical practice.

Daniel C. Dennett (Center for Cognitive Studies, Tufts University, Medford, MA 02155-5555, USA. daniel.dennett@tufts.edu) is Distinguished Arts and Sciences Professor, Professor of Philosophy, and Director of the Center for Cognitive Studies at Tufts University. After reading philosophy at Harvard he went to Oxford to work with Gilbert Ryle, under whose supervision he completed his DPhil in 1965. He taught at UC Irvine from 1965 to 1971, when he moved to Tufts, where he has taught ever since, aside from periods visiting at Harvard, Pittsburgh, Oxford, and the Ecole Normal Superieur in Paris. His many books are widely read, notably *Consciousness Explained* (1991) and *Darwin's Dangerous Idea* (1995).

K. Anders Ericsson (Department of Psychology, Florida State University, Tallahassee, Florida 32306-1270, USA. ericsson@darwin.psy.fsu.edu) is Conradi Eminent Scholar and Professor of Psychology at Florida State University. In 1976 he received his PhD in Psychology from University of Stockholm, Sweden. His research with Herbert Simon at Carnegie-Mellon University on verbal reports of thinking is summarized in a book *Protocol Analysis: Verbal Reports as Dat*a (1984/1993). Currently he studies the cognitive structure of expert performance in domains such as music, chess and sports. He is the co-editor of *Toward a General Theory of Expertise* (1991), *The Road to Excellence* (1996), and *Expert Performance In Sports* (2003).

Shaun Gallagher (Department of Philosophy, Colbourn Hall 411, University of Central Florida, Orlando, FL 32816-1352, USA. gallaghr@mail.ucf.edu) is Professor of Philosophy and Cognitive Science at the University of Central Florida. His research and teaching interests include phenomenology and philosophy of mind, cognitive science, hermeneutics, and theories of the self and personal identity. He co-edits the interdisciplinary journal *Phenomenology and the Cognitive Sciences* and has authored three books, *Hermeneutics and Education* (1992), *The Inordinance of Time* (1998) and *How the Body Shapes the Mind* (forthcoming in 2004). He has edited and co-edited several volumes, including *Models of the Self* with Jonathan Shear (1999).

Patrick Haggard (Institute of Cognitive Neuroscience, University College London, 17 Queen Square, London. WC1N 3AR. patrick@psychol.ucl.ac.uk) studied experimental

CONTRIBUTORS

psychology at Cambridge University, obtaining his PhD in 1991. He then worked at Oxford University Laboratory of Physiology before moving to a lectureship in psychology at University College, London. His research interests centre on the relationship between physiological activity in the sensorimotor areas of the brain and the conscious experiences of agency and embodiment. He is the leader of the Motor Control research group at UCL's Institute of Cognitve Neuroscience.

Anthony Jack (Washington University Campus Box 8225, 4525 Scott Avenue, St Louis, MO 63110, USA. *Email: ajack@npg.wustl.edu*) worked at Harwell Nuclear Physics Laboratories and graduated from Balliol College, Oxford, in psychology and philosophy in 1994. His PhD was in psychology at University College London, involving both experimental and philosophical/theoretical work on pattern masking and perceptual awareness. He then worked at the Institute of Cognitive Neuroscience in London, focusing largely on consciousness and its relation to executive functions of prefrontal cortex. He currently works on fMRI imaging of attention and perception at Washington University.

Helen Johnson (Institute of Cognitive Neuroscience, University College London, 17 Queen Square, London. WC1N 3AR) is a PhD student at the Institute of Cognitve Neuroscience, University College, London.

David A. Leopold (Max Planck Institut für biologische Kybernetik, Spemannstrasse 38, 72076 Tübingen, Germany. *david.leopold@tuebingen.mpg.de*) is a research scientist at the Max Planck Institute for Biological Cybernetics in Tuebingen, Germany. His research focuses on the neural mechanisms underlying visual perception in primates, which he has investigated extensively with both psychophysical and electrophysiological experiments.

Nikos K. Logothetis (Max Planck Institut für biologische Kybernetik, Spemannstrasse 38, 72076 Tübingen, Germany) is director of the department 'Physiology of Cognitive Processes' at the Max Planck Institute for Biological Cybernetics, in Tübingen. His research interests include the neurophysiology of multistable perception, object vision, and most recently functional magnetic resonance imaging (fMRI) in nonhuman primates.

Antoine Lutz (W.M. Keck Laboratory for Functional Brain Imaging and Behavior, Waisman Center, University of Wisconsin-Madison, 1500 Highland Avenue, Madison, WI 53703-2280, USA. *alutz@wisc.edu*) is a postdoctoral scientist at the W.M. Keck Laboratory. After receiving his diploma as telecom engineer (INT, Evry), he did his maîtrise (BA) in Philosophy at the University of Sorbonne, Paris IV, under the supervision of Natalie Depraz, and his PhD in Cognitive Neuroscience at the University of Jussieu, Paris VI, under the supervision of Francisco Varela. His current experimental research is focused on the neurofunctional and neurodynamical characterization of meditative states in a group of highly trained Buddhist practitioners.

Alexander Maier (Max Planck Institut für biologische Kybernetik, Spemannstrasse 38, 72076 Tübingen, Germany) is a PhD student at the Max Planck Insitutute for Biological Cybernetics. He is currently using neurophysiological techniques to investigate multistable perception, with particular regard to how visual memory contributes to our perceptual organization of ambiguous patterns.

Anthony J. Marcel (Medical Research Council Cognition and Brain Sciences Unit, 15 Chaucer Road, Cambridge CB2 2EF, UK. *tony.marcel@mrc-cbu.cam.ac.uk*) is a Senior Scientist at the Medical Research Council Cognition and Brain Sciences Unit, Cambridge. He is an Honorary Fellow in Neurology at Adden- brooke's Hospital, on the

CONTRIBUTORS

Biology B Faculty at Cambridge University, and a Professore a Contratto at the Universitá degli Studi di Padova. He currently carries out research and publishes on consciousness, bodily experience, action, space, attention, and emotion. He jointly edited *Consciousness In Contemporary Science* and *The Body and the Self.*

Josef Parnas (The Danish National Research Foundation: Center for Subjectivity Research, University of Copenhagen, Købmagergade 46, DK-1150 Copenhagen N, Denmark. *jpa@cfs.ku.dk*) is professor of psychiatry at the University of Copenhagen and the Danish National Research Foundation: Center for Subjectivity Research and a Medical Director at the University Department of Psychiatry, Hvidovre Hospital in Copenhagen. His principal research interest is aetiology and psychopathology of schizophrenia. He has numerous original publications in the domains of epidemiology and genetics, pathogenesis, risk factors, and nosological issues in schizophrenia spectrum disorders. One of his current activities concerns potential integration of continental phenomenology with empirical psychiatric research.

Gualtiero Piccinini (Dept of Philosophy, Washington University, Campus Box 1073, One Brookings Dr., St. Louis, MO 63130-4899, USA. *gpiccini@artsci.wustl.edu*) is now a postdoctoral fellow at Washington University in St. Louis, having gained his PhD from the University of Pittsburgh. He works in philosophy of mind and philosophy of science, especially on themes related to computational theories of mind and brain.

Andreas Roepstorff (PET-Centre, Aarhus University, Nørrebrogade 44, DK-8000 Aarhus C, Denmark. *Email: andreas@pet.au.dk*) received his PhD in 2002 and is assistant professor at the Centre for Functionally Integrative Neuroscience at the University of Aarhus. Educated in biology and social anthropology, he has done research in anthropology of knowledge, basic neuroscience and cognitive neuroscience.

Evan Thompson (Department of Philosophy, York University, 4700 Keele Street, Toronto, ON M3J 1P3, Canada. *evant@yorku.ca*) holds a Canada Research Chair in Cognitive Science and the Embodied Mind in the Department of Philosophy at York University, Toronto, Canada. He is the co-author with Francisco J. Varela and Eleanor Rosch of *The Embodied Mind* (1991); author of *Colour Vision* (1995); and editor of the special *JCS* collection *Between Ourselves: Second Person Issues In the Study of Consciousness* (2001). He is currently finishing a new book, begun with the late Francisco Varela, titled *Radical Embodiment: The Lived Body In Biology, Cognitive Science, and Human Experience.*

Timothy D. Wilson (Dept. of Psychology, 102 Gilmer Hall, PO Box 400400, University of Virginia, Charlottesville, VA 22904-4400, USA. *twilson@virginia.edu*) is Sherrell J. Aston Professor of Psychology at the University of Virginia, where he has taught since 1979. He is the author of *Strangers to Ourselves: Discovering the Adaptive Unconscious* (2002) and co-author of *Social Psychology* (4th ed. 2002). He has published articles in the areas of self-knowledge, the limits of introspection, the consequences of introspecting about one's attitudes, mental contamination, and affective forecasting.

Dan Zahavi (The Danish National Research Foundation: Center for Subjectivity Research, University of Copenhagen, Købmagergade 46, DK-1150 Copenhagen N, Denmark. *zahavi@cfs.ku.dk*) is professor at and Director of the Danish National Research Foundation: Center for Subjectivity Research, University of Copenhagen. He has published widely in phenomenology and philosophy of mind, and is the author and editor of numerous books including *Self-awareness and Alterity* (1999), *Exploring the Self* (2000), and *Husserl's Phenomenology* (2003).

Anthony I. Jack and Andreas Roepstorff

Why Trust the Subject?

Preliminary Remarks

It is a great pleasure to introduce this collection of papers on the use of introspective evidence in cognitive science. Our task as guest editors has been tremendously stimulating. We have received an outstanding number of contributions, in terms of quantity and quality, from academics across a wide disciplinary span, both from younger researchers and from the most experienced scholars in the field. We therefore had to redraw the plans for this project a number of times. It quickly became clear to us that the collection would expand beyond the scheduled double issue of the *Journal of Consciousness Studies*. A triple issue was then drafted, but the number of excellent contributions continued to grow. We therefore had to reconsider the publication plans again, and the decision was made to publish an extended collection of papers in discrete instalments. At present substantial progress has been made towards determining the content of a second double issue of the *Journal of Consciousness Studies*, due summer 2004. A third instalment now appears to be a real possibility. We welcome enquiries from authors interested in submitting to later instalments, especially those offering a novel perspective that is not otherwise represented. However, we do not intend continue this collection indefinitely. In putting together the first major interdisciplinary collection on this topic, we view our task as that of providing a starting point. Sufficient outlets exist to support ongoing debate.[1]

The idea for this collection first took shape when we proposed it to the managing editor of *JCS*, Anthony Freeman, at the 'Towards a Science of Consciousness' conference in Skovde, Sweden, August 2001. Since then, he has been involved in every stage of its development and construction. His editorial experience and his patient assistance have been invaluable to us and to the collection.

---

[1] In particular, authors may wish to submit to regular issues of the *Journal of Consciousness Studies*; to *Consciousness and Cognition*; and to the recently created journal *Phenomenology and the Cognitive Sciences*.

Why 'Trusting the Subject?'

In the context of cognitive science, the title of this volume *Trusting the Subject?* carries a double meaning. As touched on by Tony Marcel (this volume), the establishment of scientific knowledge is inherently bound up with notions of trust, which carries a social history of its own. From the days of the gentleman scientists in the learned societies of the seventeenth century, via the Introspectionists at the turn of the twentieth century, and the Behaviourists that dominated psychology throughout the middle part of the twentieth century, to the cognitive neuroscientists who have recently begun to transform psychology, differing understandings of trust both in the experimenter and in the experimental subject have served as semi-stable foundations for the generation of facts and the establishment of knowledge about the human mind. In doing cognitive science and consciousness research, two different levels of trust are therefore at stake. On one level it refers to the interaction between two concrete persons, an experimenter and a volunteering experimental subject, and the extent to which the former uses the reports of the latter as some sort of evidence for scientific inquiry. On another, more general level, however, the notion of 'subject' refers to the actual scientific enterprise of inquiring into the mind. Can one trust the subject of cognitive research, the human mind? Scientists are used to relying on instruments that they themselves have manufactured — technologies whose mode of operation, and limitations, are usually well understood. The unique challenge facing a science of consciousness is that that the best instrument available for measuring experience depends on cognitive processes internal to the subject. So just how much faith can we place in the capacity of the mind to understand itself? In principle, the construction of a maximally robust methodology for introspective evidence would require a detailed understanding of the operation of introspective processes — the processes that mediate the acquisition of introspective knowledge and underlie the production of introspective reports (Jack & Shallice, 2001). Given that knowledge, absolute trust in introspective evidence could be warranted. The practical question is: What attitude should we take given our relative ignorance of introspective processes?

Most scientists do not have, or at least cannot coherently formulate, any principled objection to introspective reports; rather, they simply lack faith that introspective reports are reliable in practice. Some find support for this belief in the informal interviews they conduct at the end of experiments. Subjects, it appears, frequently vary in their take on the experiment. Yet experimenters should not be surprised that such undisciplined reports do not provide consistent results. Like all methodologies, introspective methods require a number of factors to be controlled: When is the subject attending to their experience? To what aspect of experience are they attending? What 'model' are they using to interpret and filter their experiences? In some cases accurate reports will require at least some minimal training, to provide subjects with concepts they can use to effectively communicate their experience.

Despite such widespread pessimism, there are numerous reasons to believe in the accuracy of introspective reports. First and foremost, could any normally socially functioning human seriously doubt that they have succeeded, at least on occasion, in accurately accessing information about their own internal states, for instance concerning their: emotions, state of concentration, thoughts, actions, cognitive strategies, confidence, imagery and focus of attention? Second, numerous experiments, notably in psychophysics, memory and problem solving, illustrate the reliability of reports when they are carefully collected. Third, an informal reliance on introspective evidence is *ubiquitous* in psychology and cognitive science. It generates many of the hypotheses that psychologists seek to test using objective sources of evidence, it underlies their understanding of cognitive tasks or 'task analysis', and it frequently informs the questions and objections they offer as referees. Introspective understanding even forms the basis of many of the categories used to describe branches of psychological research (e.g. 'attention', 'episodic memory', 'awareness'). If psychologists are reliant on introspection as a source of anecdotal evidence, then shouldn't scientific instinct suggest that a more formal, disciplined and systematic treatment of the evidence will prove more productive? At the very least, we should like to clearly understand what would limit this strategy.

A common, and historically motivated, misconception of introspective methodology views it as in competition with 'objective' (behavioural and neural) methods. In contrast to this, we have argued that the interpersonal perspective involved in the communication of experience is already an integral part of standard methodology in cognitive science (Jack & Roepstorff, 2002; Roepstorff & Frith, in press). It is reflected both by the experimenter's attempts to offer the subject a model for how they should carry out the experimental task (the task instructions or 'script'); and again when the subjects attempt to communicate their actual experience of the task, typically elicited in the informal post-hoc interview that is considered good experimental practice. Both the experimenter's model of what the task involves, and the reports elicited from subjects, frequently serve to inform the interpretation of cognitive experiments. Our aim is therefore to expand and improve upon current practice, through the explicit and formal recognition of the larger framework of 'script-report' that encompasses the standard formalisation of 'stimulus-response' in behavioural methods. The advantages of acknowledging this larger framework, and formalising new methods for capitalising on it, cannot be accomplished unless we also maintain attention to the behavioural factors that allow for tight experimental control and inference to underlying mechanism. In our view, 'stand-alone' introspective methods and 'armchair' introspection are not likely to carry us very far. Hence our emphasis on 'triangulation' — the use of introspective, behavioural, and physiological evidence in concert (Jack & Roepstorff, 2002) — and the specific emphasis of this collection on the role of introspective evidence in cognitive science.

The Validity of Introspective Evidence

It is important to realize that no principled problem stands in the way of the scientific assessment of various types of introspective evidence. The testing of the reliability, consistency and validity of various types of introspective report measures lies well within the orbit of currently available methods.

A measure[2] may be called 'reliable'[3] if it yields the same results when tested in multiple sessions over time ('test–retest reliability') and across individuals (a cousin of 'inter-rater' and 'inter-observer' reliability). Of course, subjects' reports may differ, and so appear to be unreliable, simply because their internal mental processes and states vary. Thus it is critical to establish well controlled experimental conditions for eliciting reports. The considerable advances in behavioural science since the time of the Introspectionists offers experimenters considerable advantages in this regard (see Ericsson, this volume). Not only do these advances make it much more probable that experimenters can establish conditions under which introspective measures can be shown to be reliable, they also provide much greater insight into the behavioural and neural correlates of experiential phenomena.

A measure may be called 'consistent' when it can be shown that the results are not due to specific features of the measurement technique. Tests of consistency provide a means of checking that the observed effect is not due to a methodological artefact. Thus we might test the consistency of introspective evidence by comparing immediate forced-choice button-press reports with retrospective and open-ended verbal reports. In this way we might establish, for instance: that the results of forced-choice button-press reports have not been influenced by variations in the criterion for response or by automatisation of response such that they no longer constitute true introspective reports; and that retrospective reports have not been distorted by forgetting or memory interference effects.

'Validity' is the most important factor to establish, yet it is also the most theoretically complex, and a particularly vexed issue in cognitive science. A measure is validated when it can be shown to accurately reflect the phenomenon it purports to measure. Validity is complex because scientific measures are often simultaneously interpreted as providing evidence for phenomena at a number of different levels. A rough characterisation of three major sources of evidence in cognitive science might read as follows:

> Data from functional Magnetic Resonance Imaging (fMRI) serves most directly as evidence of cerebral blood flow (which has been validated), less directly as evidence for neural activity (which is in the process of being properly validated), and least directly as a means of identifying and localising specific cognitive functions (far from well validated).

---

[2] The term 'measure' is used in a very general sense here, and is not meant to imply the ability to map the underlying process in any specific manner (e.g. using a continuous scale). Thus a 'measure' might just be an experimental method for identifying the presence or absence of a particular internal state.

[3] Epistemologists often use the term 'reliability' to refer to the accuracy and/or validity of a particular knowledge source. It has a slightly different meaning in scientific contexts.

Behavioural measures (e.g. the averaging of reaction time measures over multiple trials) serve most directly as evidence for stable patterns of behaviour, less directly as a means of assessing information processing, and least directly as means of establishing the existence and operation of specific cognitive functions.

Introspective reports serve most directly as evidence about the beliefs that subjects have about their own experience, less directly as evidence concerning the existence of experiential phenomena, and least directly as evidence concerning the operation of specific cognitive functions.

The issue of validity is particularly vexed in cognitive science because there has been a long history of theoretical and philosophical disagreements about the nature of the mental — about what psychological evidence is ultimately serving as evidence for. The early psychologists regarded consciousness as the mark of the mental (Wilkes, 1988). Thus it is often said that scientific psychology began with 'psychophysics' — a project initially conceived by Fechner and Weber as an attempt to find law-like relationships between the physical properties of the stimulus and the experiential properties of the percept. Specifically the Weber-Fechner law was put forward to describe the relationship between physical intensity and 'felt' intensity. In stark contrast, the Behaviourists rejected any reference to internal mental states, and defined the purpose of psychology as that of identifying stable patterns of behaviour. This eventually gave way to a growing sense that behaviourist science (e.g. the description of 'processes' such as habituation, classical conditioning, overshadowing, etc.) served primarily as a means of re-describing the data. Information processing accounts provided a well grounded way of making inferences from this data to internal processes and states. Yet the dominance of the information processing model, and in particular its strong emphasis on behavioural performance, has sometimes made it difficult for other sources of evidence to find a purchase. Information processing accounts are primarily concerned with what subjects are capable of doing with the information in the stimulus, as indicated by the appropriateness of their behaviour for achieving a specific goal (whether that be a sub-personal goal, such as making an accurate visual saccade, or a personal goal, such as achieving good performance in a logical reasoning task). The observation that particular parts of the brain are preferentially involved in different tasks appeared, at least initially to many psychologists, to have little direct relevance to understanding cognition. Similarly, introspective reports do not, at least at first, appear to provide data relevant to information processing accounts. The goal of introspective report is to provide an accurate description of experience. Since the experimenter cannot directly observe subjects' experiences, there is no easy way to assess the accuracy of their performance. Putting the point another way, without knowing what information subjects have internal access to, psychologists can't use introspective reports in the same way they use objective behavioural measures to aid in the construction of information processing models. Thus the publicly inaccessible nature of experience can seem to militate against the validity of introspective evidence.

Introspective reports cannot be treated in the same way as other behavioural measures, yet this does not preclude their use to inform information processing accounts. The expanding view afforded by the increasing influence of neurophysiological evidence provides greater opportunity for introspective evidence to find other points of purchase. A consequence of this is that results previously thought to indicate the unreliable nature of introspective reports may now be seen in a different light. A good example comes from the historical account provided by Anders Ericsson (this volume), as follows:

> [A] large body of research has attempted to relate the level of accurate recall of a presented picture to the reported vividness of the memory (McKelvie, 1995; Richardson, 1988). To everyone's surprise, no clear relation between the amount of accurately recalled information and reported vividness has been found. Participants who reported recalling a presented stimulus as vividly and clearly as if it remained visible did not recall more accurate information than those who reported diffuse memory images. These and other puzzling findings, such as the reported persistence of visual eidetic images (Haber, 1979), confirmed the opinions of many experimental psychologists that introspective judgments about experience were frequently misleading and inconsistent with measures of performance (p. 6).

Ericsson is describing the sort of thinking that has led psychologists to conclude that introspective reports are invalid. Let us carefully consider the logic of this conclusion. What the research shows is that the intuitively appealing idea that memory accuracy and reported vividness should correlate turns out to be wrong. Yet the conclusion that the reports are invalid depends on how you interpret them. If the reports are interpreted as being reports about memory accuracy, then clearly they are invalid. Experimenters should not trust reported vividness as a guide to memory accuracy.

However, we might approach these reports in another way. Instead of attempting a direct translation of these reports into information processing terms (i.e. as describing the efficiency and thus accuracy of the processes underlying recall) we might more literally construe them as 'introspective judgments about experience'. According to this strategy, we should remain agnostic, at least for the time being, about the correspondence between experience and information processing. This gives us a three way relationship. We have memory accuracy, experienced vividness, and reported vividness. Given this framework, we can see that at least one of the two relationships must break down. Either memory accuracy does not correspond to experienced vividness, or experienced vividness does not correspond to reported vividness. Further, we can see that two separate suppositions support these different relationships. The first relationship is supported by a folk-psychological belief, the belief that perceptually vivid memories should be more accurate than diffuse non-vivid memories. The second relationship is supported by the view that the reports in question are accurate and valid. Given this framework we can see that one strong possibility is that the reports are accurate but that the folk-psychological belief is false. Furthermore, we can seek evidence that would provide some support for this view. For instance, it is reasonable to hypothesize that activity in visual cortical areas will correlate with the experienced vividness of memories (Wheeler *et al.*, 2000). Thus, evidence of a correlation between reported vividness and visual activity would support the view that the reports are valid whilst the folk-psychological belief is false. We have demonstrated a closely related result when subjects are asked to provide immediate reports of difficult to perceive (masked) visual stimuli. Summerfield *et al.* (2002) used EEG to show that gamma band activity over occipital (visual) cortex correlates with reported vividness, even for stimuli that were *incorrectly* identified.

Ericsson's example, and other similar cases,[4] illustrate that part of the reluctance of psychologists to ascribe validity to introspective report measures derives from a tendency that might be called 'the rush to operationalize'. For historical reasons, deriving from the positivism of the behaviourists, experimental psychologists are highly reluctant to adopt the strategy of interpreting introspective reports in the most straightforward and direct manner, as telling us about experience. Instead they seek what are called 'operational definitions' — they seek to define internal states and processes in terms of their behavioural effects. This emphasis on operational definitions ensures that the claims that psychologists make are concrete, specific and falsifiable. Yet the problem with adopting this strategy when interpreting introspective reports is obvious: despite the prevalence of folk-psychological beliefs, the true relationship between experience and behaviour is often difficult to ascertain. When scientists are forced to make a choice between appearing somewhat vague on the one hand and relying on an untested and intuitive assumption on the other, scientific progress is often better served by temporarily maintaining a degree of vagueness.

The reluctance of psychologists to interpret introspective reports as telling us about experience is also shared by the philosopher Daniel Dennett (this volume). We regard Dennett's work as important for a number of reasons: First because he has long been at the forefront of a movement to discuss experiential phenomena in cognitive science and encourage debate concerning their interpretation. Second because he has formulated an explicit position concerning the scientific use of introspective evidence, which he calls 'heterophenomenology'. Third, because in our view, his position provides the best representation of the underlying philosophy that guides current practice in cognitive science. Dennett (this volume) argues that scientists should only go so far as to make claims about the beliefs that subjects have about their experiences. Scientists should stop short or 'reserve judgment' about the truth of these claims. We agree that this approach, which Dennett calls the 'bracketing' of experience, must play a key role in the establishment of a methodology for introspective evidence. Only by remaining

---

[4] Jack & Shallice (2001) and Jack (2001) discusses a similar mistake made by researchers in the field of perception without awareness. Until recently, forced-choice discrimination performance was seen as the gold-standard 'objective' measure of awareness, and the lack of correspondence with introspective reports was interpreted as illustrating their invalidity. After 50 years, researchers recently realized that the lack of correspondence was not due to problems with the introspective reports, but to contamination of the 'objective measure of awareness' by non-conscious information. Currently favored behavioral measures of awareness, such as those based on the Jacoby Process Dissociation Procedure (Jacoby, 1991), correspond much more closely with introspective reports of awareness.

neutral about the accuracy of any particular introspective report, can we critically assess the evidence concerning the reliability, consistency and validity of different types of report. Dennett's cautious approach thus helps to avoid the trap of over-interpreting introspective evidence — the problem of undue trust. Dennett's view is closely related to that espoused by many (but by no means all) psychologists: they are willing to acknowledge the role of introspective evidence in generating hypotheses and in influencing preliminary interpretations of results, yet they view the final arbiter, and the real business of science, as lying in the collection of objective evidence.

We believe that cognitive science can do better than this, and we are sceptical of certain aspects of Dennett's position. Specifically, Dennett appears to insist that we must *always* reserve judgment about the veracity of subjects' beliefs about their experiences, pending verification of their claims using objective evidence. Our view differs in two ways. First, we do not believe that it is possible to use objective evidence to directly test or 'verify' the accuracy of subjects' reports. Second, we do not find motivation for the claim that it is *always* necessary to reserve judgement about the accuracy of introspective reports. Instead we take the view that we should place a degree of trust in introspective reports, proportional to the evidence of their validity.

Again, both these issues come down to how the introspective evidence is interpreted. When introspective evidence is interpreted as evidence about the operation of cognitive processes, such as a claim about information processing accuracy, then it is clear that objective evidence can be used to directly test its validity. Thus Ericsson's example illustrates how reported vividness was found to be an invalid measure of memory accuracy. Alternatively, in their perceptual masking experiment, Summerfield *et al.* (2002) found that reported vividness consistently correlated well with discrimination accuracy across a range of masking times, except when masking time was very short, at which point the correlation between reported vividness and discrimination accuracy broke down.

When introspective evidence is interpreted more directly, as evidence about the subject's experience, objective measures can no longer serve to directly test their validity. In this case, as the discussion of Ericsson's example above illustrates, it may be possible to find convergent evidence that lends support to the view that the reports are valid. Yet objective evidence cannot be used to *directly* verify or falsify the accuracy of introspective reports about experience. This does not mean that reports are not falsifiable. It means falsification can only be achieved indirectly, by means of inference to the best explanation. For example, it may be that subjects who are suddenly placed in highly exasperating situations have an initial tendency to deny that they are angry. This might happen because the onset of anger causes attention to focus exclusively on the perceived source of irritation, so diverting attention from inner states and preventing accurate self-ascription. In this case, the falsity of the subjects' reports might be established by a convergence of evidence: the inconsistency of concurrent reports with later retrospective reports, the presence of behavioural indicators of anger, and other evidence that supports the proposed hypothesis about the effects of anger on attention.

Dennett's heterophenomenological perspective differs from ours because he does not recognize the same gap between experience and behaviour. His philosophical position identifies mental states with patterns of behaviour — just as many psychologists are apt to do in practice. In our view patterns of behaviour, neural processes and experience exist as distinct facets of the mental. Thus we maintain that objective measures can only provide tangential evidence about experience, by means of an underlying theory of mental processes. Introspective evidence provides the most direct view of the experiential facet of mental processes. Just as establishing the reliability and consistency of behavioural measures assures a reasonable degree of trust in their validity as measures of information processing; so establishing the reliability and consistency of introspective reports should assure a reasonable degree of trust in their validity as measures of experience. We take it to be obvious that introspective evidence, and only introspective evidence, has 'face validity' in the measurement of experience. No doubt introspective reports will sometimes be mistaken, and this may be established by convergent evidence, yet the balancing of equivocal evidence should always be weighted in favour of introspective reports.

We take this point to be important to establish because a degree of trust is actually essential to scientific progress. Although some degree of scepticism is always advisable, undue scepticism prevents the bold hypotheses that push science forward. The scientist who never dares to trust her methods will, of course, never allow herself to discover a thing. Ericsson's psychologists, who were so quick to interpret the reports of vividness as invalid, would never find the motivation to look for other correlates of those reports. So long as cognitive science continues to doubt the face validity of introspective reports, it will never conduct the investigations necessary to provide full validation of those measures; nor can it ever hope to provide scientific accounts of experience. We shall need to take the time to explore and understand experience, before we can hope to generate strong hypotheses about its behavioural and neural correlates.

If this view is correct, then it will have profound implications for methodology in cognitive science. At present the experimental assessment of awareness, in fields such as implicit learning, memory and perception without awareness, is largely achieved by means of objective performance measures. Our view certainly does not preclude the use of objective measures of awareness; however it does turn the current line of thinking about validation, expressed in Dennett's philosophy, on its head. Where experiential phenomena are concerned, it is *objective* measures that must seek validation by establishing their correspondence with *introspective* measures, and not vice versa. Furthermore, psychologists should be willing to accept the value of investigations that focus primarily on data from introspective reports, provided the cognitive tasks employed are also well controlled. At present few mainstream psychological journals would accept such investigations for publication. Instead, at present, they accept experiments that purport to use objective evidence to substantiate claims about experience.

The Significance of Introspective Evidence

Why should we care about introspective evidence? What can it tell us, and how can it benefit cognitive science? We will discuss three advantages of introspective evidence.

(1) Understanding mechanism

Introspective evidence may assist in the normal business of cognitive science, as an additional source of evidence that can inform and guide mechanistic accounts of mental function. To motivate this view, we need only make a few minimal and plausible assumptions about introspection. First, we assume that introspective processes have access to some limited subset of the functional properties of mental states (Jack & Shallice, 2001). Second, we assume that introspective processes are capable of performing some basic information processing operations on this information, which may be understood by analogy to perception:

(i) We have a capacity to learn to recognize internal states that have occurred on a number of previous occasions, such that we are able to recognise further recurrences of those states (Siegler & Stern, 1998), given that we are attending internally.
(ii) We have a capacity both to encode and to recall information about previous internal states, so allowing us to make comparisons between states.
(iii) We are able to attend selectively to specific features of internal states, so allowing us to compare states along a number of dimensions. For example 'This headache is sharper than the one I had yesterday, it was duller. The headache yesterday was throbbing, the one today is continuous.'

Finally, we assume that we evolved our capacities to recognize and distinguish between our own internal states, perhaps by the extension of existing perceptual and mnemonic processes (Jack, 2001), and that these capacities serve a useful function. In order to support an evolutionary advantage, we suppose that introspective processes are at least reasonably successful at recognizing and discriminating between internal states.

If these assumptions are correct, then there are straightforward ways in which we can use introspective reports to provide clues about functional differences between mental states. A strong example of this comes from work on synaesthesia (Cytowic, 1997). Synaesthetes report that particular sorts of experiences (e.g. hearing a word) are similar along a certain dimension to other, normally quite different sorts of experience (e.g. heard words have the colour properties of visual experience). This has led to hypotheses about the functional similarities of the two states, which have been borne out by neural tests (Paulesu et al., 1995).

Another more general way in which introspective evidence may prove useful to cognitive science concerns the strategy that psychologists use to identify putative cognitive processes pending further investigation. It is common practice for psychologists to identify processes by reference to a particular behavioural paradigm, good performance on which is hypothesized to require the process in question. One problem with this strategy of defining processes in terms of behavioural tasks is that psychologists are then apt to generalize on the basis of the task. They assume that similar tasks should evoke similar processes, when this is often not the case. For example, psychologists frequently talk about 'recognition memory' and 'working memory' in a manner that suggests they are referring to an actual cognitive process. Yet there is now a wealth of evidence that both working memory tasks and recognition memory tasks involve a range of different cognitive processes (which have been shown to be engaged to differing extents depending on manipulations of the task, e.g. Chein & Fiez, 2001; Mandler, 1980; Rowe & Passingham, 2001). As a result, the only presently salvageable notion of 'working memory' as a cognitive process is so vague as to have almost no explanatory value. In contrast 'episodic memory', a construct that is almost unique in psychology in that it was explicitly defined by Endel Tulving (1972) in phenomenological terms, appears to be maintaining its explanatory value very well. Given the clear phenomenology associated with different aspects of working memory tasks (e.g. imagery, sub-vocal rehearsal, 'refreshing' of information, 'chunking', 'holding information in mind', etc.) one cannot help but wonder whether research in this area would have fared better if its constructs had been defined at the outset in phenomenological terms. In general, it seems that higher cognitive processes such as those involved in working memory tasks, tests of executive function and problem solving, are particularly amenable to analysis using introspective reports and verbal protocols (Jack & Roepstorff, 2002). Thus Tulving's strategy may prove particularly productive for investigations of these processes.

(2) Understanding consciousness

The pursuit of scientific theories of consciousness has clearly become a hot topic in cognitive science. Yet, surprisingly, few researchers have explicitly acknowledged the central importance of understanding introspection to this enterprise. When we have pushed consciousness researchers on the need to produce some sort of account of the mechanisms that allow us to acquire knowledge about experience, many have replied that such higher-order processes are not their primary interest. They claim to be interested in the nature of experience itself ('first-order awareness' or 'phenomenal consciousness'), not the processes that allow us to make judgments and reports concerning our experiences ('second- order awareness' or 'reflexive consciousness'). These researchers have simply missed the point. From an epistemological point of view, introspection is the *sine qua non* of consciousness. Without introspection, we simply wouldn't know about the existence of experience. And without a good theory of introspection, we have no way of establishing what sorts of claims about experience are justified.

Good scientific theories succeed because they make sense of the data. A theory of consciousness must not only make sense of the neural and behavioural data — it must also make sense of experience. Yet it is important to realize that

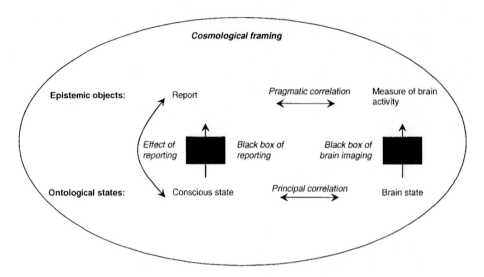

making sense of the data on experience will not always mean following the most intuitive and obvious interpretation of that data. A strong theory isn't likely to fit comfortably with all our intuitions about experience. Part of the promise of a strong theory of consciousness is that it should cause us to view our experiences in a new light.

Consider the logic behind attempts to find the neural correlates of consciousness (NCC). Studies of the NCC seek to establish a correspondence between the subject's experience and their brain state. To provide evidence for this, the researcher demonstrates a correspondence between the neural measure used (here we consider brain imaging) and the introspective reports elicited by subjects. We can understand these measures as *epistemic objects*. That is, these measures derive from the application of a set of methods and criteria that ensure some kind of validity. Within science studies, it has become customary to describe this process as 'black boxing'. This means that when the process is running smoothly 'one need focus only on its inputs and outputs, and not on its internal complexity' (Latour, 1999). Yet, as has been demonstrated by a whole range of science studies, the relationship between the resulting epistemic objects and the underlying states is in no way trivial. The 'black boxes' involve complex transformations that do not always succeed in achieving a smooth translation. The commonly applied counter strategy to this problem is to 'open the black box' in order to follow in minute details the actual transformations, reductions and amplifications involved in settling the epistemic objects.

Although it may not be generally known outside the brain imaging community, it is relatively uncontroversial that the colourful pictures of brain activity obtained by PET, fMRI, MEG or EEG are very far from realistic photographs of the brain. They are rather to be seen as complicated graphs, the outcome of a set of mathematical procedures and transformations, that could have been done differently (Roepstorff & Gjedde, 2003). To complicate things even further, there are serious discussions in the brain imaging field about what the relationship is between the largely metabolic and circulatory measures obtained and the actual behaviour of neurons. These discussions occur at two levels. They are a matter of settling the link between, for instance, fMRI measurements of the BOLD signal or PET measurements of blood flow or oxygen consumption and the underlying neuronal activity. More fundamentally, however, there is no agreement as to what should count as a proper description of brain states — should they, for instance, be identified by synaptic processing or by the firing of individual neurons? This means that our understanding of the link between the measure of brain activity and the putative brain state is constantly evolving.

A science of consciousness will also require us to address the link between reports and conscious states. We must recognize that introspective reports do not represent a transparent reflection of inner experience, but are instead the products of a complex 'black box' set of processes. There will be times when we shall need to 'open the black box' of introspective processes, before we can hope to generate a stable view of the nature of experience. One line of evidence for instabilities in our present view of experience is demonstrated by an interesting set of experiments due to Tony Marcel (1993). They show that even in a very simple psychophysical setting, the measurement of experience depends on the actual method of reporting, be that button pressing, verbal account or eye blink (see also Marcel, this volume). The relationship between the reports on the one hand, and the putative underlying conscious states on the other, is not trivial. As with the relation between brain images and brain states, it is the result of the application of particular epistemic technologies, and only through a careful interplay between black-boxing and opening the box will it become possible to elucidate this relation.

(3) Types of psychological explanation

In our view, by far the most significant role that introspective evidence can play is that of elucidating the links between different types of psychological explanation. At present, psychology is a highly fractured discipline that lacks any sound and over-arching theoretical framework. On the one hand much of psychology, in particular those areas that now come under the umbrella of 'cognitive science', is concerned with providing mechanistic accounts of mental processes. On the other hand, many branches of psychology, in particular social psychology and the therapeutic branches, are concerned with giving accounts that work at the personal, experiential, level of explanation. Thus the challenge of relating introspective evidence to objective evidence directly reflects a key challenge for psychology — that of resolving the tension between different types of psychological explanation, so finally unifying the discipline.

Timothy Wilson (this volume), whose earlier work set the standard for a generation of work on verbal reports (Nisbett & Wilson, 1977), notes that despite the frequent attacks on introspective methods, they are used successfully in many areas of psychology. Nonetheless, it is clear that introspective methods have played a far greater role in areas of psychology that might be considered branches of 'social science' by hard-nosed cognitive scientists. It seems that many cognitive scientists fear their scientific credibility would be threatened by introspective reports.

How should we understand the difference between these two types of psychological explanation? In our experience, most people have little trouble understanding what it means to provide mechanistic accounts. This is the dominant model of explanation in psychology, borrowed from the hard sciences. Yet many people fail to recognize the critical differences between these mechanistic accounts and accounts that work at the personal level. Personal-level accounts are accounts that we can make sense of on our own terms, rather than from a removed third-person perspective. Personal-level accounts help us to make sense of our experience, they inform our conscious strategies, they alter our interpersonal perceptions, and they help us to understand the implications of mechanistic accounts for our everyday lives.

A simple example of a scientific finding that can be understood at the personal level comes from the work on vividness and memory recall discussed in the previous section. The finding that the perceptual vividness of recalled information does not serve as a good guide to memory accuracy is something that we can make sense of a personal level. More than that, it is a finding that we can actively make use of at a personal level. We can use this information to inform the cognitive strategies that we employ to check the veracity of our own memories, by altering the criteria we use to ascribe confidence. Meta-cognitive strategies of this sort have been shown to influence performance in the lab, specifically on free report tasks (Koriat & Goldsmith, 1996). Increasingly, cognitive psychologists are coming to realize that a great deal of the variation observed both in long-term and working memory performance can be accounted for by the use of more or less sophisticated meta-cognitive strategies — personal first-person knowledge about how best to encode, retrieve and assess the accuracy of memories (Ericsson, 2003).

Our experiences, and the ways we think about them, are far from epiphenomenal. They make a major contribution to performance on all but the most simplified cognitive tasks, they influence our life decisions, and they directly affect our sense of well-being. By better characterizing the information we have internally available to us, and the ways in which we categorise and process that information, we may greatly improve our understanding of meta-cognitive and self-regulatory processes, and so find ways to improve them.

If many cognitive scientists find it hard to recognise personal level psychological accounts, then even more overlook their central significance for cognitive science. Scientific enquiry serves two basic purposes:

(i) As a means of understanding — the pursuit of knowledge for its own sake.
(ii) As a means of intervention — the generation of technologies and methods that allow us to influence the world.

First, let us consider science as a means of understanding. The question we need to ask ourselves is this: What is involved in understanding the mind? If we were to possess a complete mechanistic account of brain function, would we have a full and complete understanding of the mind? We find this claim implausible. Surely in order to claim that we understand the mind, we must be able to understand our own minds. For many philosophers, this just is the 'problem of consciousness'. According to this perspective, consciousness does not represent a specific and tractable issue for scientific investigation. Rather it represents the diagnosis of a serious failure in the whole discipline. The basic argument of these philosophers is clear: scientific accounts leave something out — they leave out experience. No wonder that this charge should sting. How can we expect to smuggle experience in through the back door, when we are so reluctant to collect systematic data on it? Cognitive scientists spend a great deal of time discussing the interpretation of objective evidence, yet we have never read nor heard a cognitive scientist suggest that further work is needed to understand what it is like to carry out an experimental task.

Understanding personal-level explanations does not merely constitute the greatest intellectual challenge facing cognitive science. There are also eminently practical reasons for wanting to develop better personal-level accounts and seeking to understand their points of contact with mechanistic accounts. It is the personal level that we primarily care about. With chronically ill patients, it is the level of pain they experience that concerns us. Only by extension do we concern ourselves with their galvanic skin response, their cortisol levels, or their neural activity.

Second, let us consider science as a means of intervention. How might the science of the mind intervene to improve our lives? One of the most interesting features of accounts that work at the personal level is that they can serve to directly alter mental function. For instance, cognitive therapy is a method that works simply by encouraging subjects to observe their own experiences and to think about them in different ways. It is a highly effective method. It has long been a treatment of choice for anxiety-related disorders, and it has recently been shown to be highly effective in preventing relapses in depression (Teasdale *et al.*, 2002).

More generally it is clear that the population at large has a powerful hunger for interventions that work at the personal level. Walk into any major bookshop, and you will find that popular psychology makes up one of the largest sections. Personal-level explanations and training protocols represent both the most humane and the most publicly acceptable means of intervention that any science of the mental can hope to offer. So shouldn't a major goal for cognitive science be the use of modern scientific methods to better inform our understanding of the processes that mediate the influence of personal-level explanations?

It is unclear how much effort cognitive science is likely to put into understanding interventions at the personal level. However, it is abundantly clear that the mechanisms of science funding are in place to ensure a vast increase in the number of interventions available at the genetic, neural and pharmacological levels. Such progress will bring with it a clear and troubling concern: how will we be able to understand the effects of these interventions? Specifically, how can we hope to attain true informed consent from people undergoing such interventions, unless we are able to explain their effects at the personal level? It seems unlikely that explaining the neural and behavioural consequences of interventions in brain function will prove adequate. If we are to pursue a truly ethical course, we shall surely need to do our best to explain to patients how these interventions may alter their concept of self — how they will alter the ways in which they experience their everyday life. If we continue to refuse to trust the subject, the subject will have no reason to trust us. Cognitive scientists should not fear that introspective evidence will impugn the scientific credibility of their work. They should fear the Frankenstein science they will create without it.

References

Chein, J.M. & Fiez, J.A. (2001), 'Dissociation of verbal working memory system components using a delayed serial recall task', *Cereb Cortex*, **11** (11), pp. 1003–14.

Cytowic, R.E. (1997), 'Synaesthesia: phenomenology and neuropsychology', in *Synaethesia*, ed. S. Baron-Cohen & J.E. Harrison (Oxford: Blackwell).

Ericsson, K.A. (2003), 'Exceptional memorizers: Made, not born', *Trends in Cognitive Sciences*, **7** (6), pp. 233–5.

Jack, A.I. (2001), 'Paradigm Lost: A review of *Consciousness Lost and Found* by Lawrence Weiskrantz', *Mind & Language*, **16** (1), pp. 101–7.

Jack, A.I. & Roepstorff, A. (2002), 'Introspection and cognitive brain mapping: From stimulus-response to script-report', *Trends in Cognitive Sciences*, **6** (8), pp. 333–9.

Jack, A.I. & Shallice, T. (2001), 'Introspective physicalism as an approach to the science of consciousness', *Cognition*, **79** (1–2), pp. 161–96.

Jacoby, L.L. (1991), 'A process dissociation framework: Separating automatic from intentional uses of memory', *Journal of Memory and Language*, **30** (5), pp. 513–41.

Koriat, A. & Goldsmith, M. (1996), 'Monitoring and control processes in the strategic regulation of memory accuracy', *Psychological Review*, **103** (3), pp. 490–517.

Latour, B. (1999), *Pandora's Hope: Essays on the Reality of Science Studies* (Cambridge, MA: Harvard University Press).

Mandler, G. (1980), 'Recognising: The judgment of previous occurrence', *Psychological Review*, **87**, pp. 252–71.

Marcel, A. (1993), 'Slippage in the unity of consciousness', in *Experimental and Theoretical Studies of Consciousness*, ed. G.R. Block & J. Marsh (Chichester: Wiley).

Nisbett, R.E. & Wilson, T.D. (1977), 'Telling more than we can know: Verbal reports on mental processes', *Psychological Review*, **75**, pp. 522–36.

Paulesu, E., Harrison, J., Baron-Cohen, S., Watson, J., Goldstein, L., Heather, J., Frackowiac, R. & Frith, C. (1995), 'The physiology of coloured hearing', *Brain*, **118**, pp. 671–6.

Roepstorff, A. & Frith, C.D. (in press), 'What's at the top in the top-down control of action?', *Psychological Research*.

Roepstorff, A. & Gjedde, A. (2003), 'Subjectivity as a variable in brain imaging experiments' [in Danish], in *Subjektivitet og videnskab. Bevidsthedsforskning i det 21. århundrede*, ed. D. Zahavi & G. Christensen (Roskilde: Roskilde Universitetsforlag).

Rowe, J.B. & Passingham, R.E. (2001), 'Working memory for location and time: Activity in prefrontal area 46 relates to selection rather than maintenance in memory', *Neuroimage*, **14** (1 Pt 1), pp. 77–86.

Siegler, R.S. & Stern, E. (1998), 'Concious and unconscious strategy discoveries: A microgentic analysis', *Journal of Experimental Psychology: General*, **127** (4), pp. 377–97.

Summerfield, C., Jack, A.I. & Burgess, A.P. (2002), 'Induced gamma activity is associated with conscious awareness of pattern masked nouns', *Int J Psychophysiol*, **44** (2), pp. 93–100.

Teasdale, J.D., Moore, R.G., Hayhurst, H., Pope, M., Williams, S. & Segal, Z. V. (2002), 'Metacognitive awareness and prevention of relapse in depression: Empirical evidence', *J Consult Clin Psychol*, **70** (2), pp. 275–87.

Tulving, E. (1972), 'Episodic and semantic memory', in *Organization of Memory*, ed. E. Tulving & W. Donaldson (New York: Academic Press).

Wheeler, M.E., Petersen, S.E. & Buckner, R.L. (2000), 'Memory's echo: Vivid remembering reactivates sensory-specific cortex', *Proc Natl Acad Sci U S A*, **97** (20), pp. 11125–9.

Wilkes, K.V. (1988), '——, yishi, duh, um, and consciousness', in *Consciousness In Contemporary Science*, ed. A.J. Marcel & E. Bisiach (Oxford: Clarendon Press/Oxford University Press).

K. Anders Ericsson

Valid and Non-Reactive Verbalization of Thoughts During Performance of Tasks

Towards a Solution to the Central Problems of Introspection as a Source of Scientific Data

Recent proposals for a return to introspective methods make it necessary to review the central problems that led psychologists to abandon those methods as sources of scientific data in the early twentieth century. These problems and other related challenges to verbal reports collected during the cognitive revolution during the 1960s and 1970s were discussed in Ericsson and Simon's (1980; 1993) proposal for a theoretically motivated procedure to elicit valid and non- reactive concurrent verbalization of thoughts while subjects were performing tasks. The same proposal explains why other verbal reports, such as introspections, detailed descriptions or explanations, require additional cognitive activity that often leads to reactivity and invalid reports. Finally, a new proposal is sketched for how the generation of introspective reports might be incorporated within a framework for non-reactive and valid verbalization of thoughts.

The contributions to this special issue demonstrate rapidly growing interest in introspective methods for studying experience and cognitive phenomena. As long as the procedures for collecting evidence are consistent with existing cognitive theories, scrutiny of them is often limited. The real test of the validity of the data-collection procedures comes when the uncovered findings are in conflict with prevailing theories and investigators with opposing views attempt to replicate each other's findings. This paper argues that it is necessary to understand the fierce controversies over introspective evidence in order to allow the development of improved methods that will meet the standards for reproducible scientific data.

During the early twentieth century introspective data on 'imageless thought' was presented by Karl Bühler (1907) to refute the dominant theory of Wilhelm Wundt, one of the founders of laboratory research in psychology. During the subsequent fierce argument, introspective data failed to provide reproducible objective evidence that could resolve the argument. Following this and other related failures, the introspective method with its trust in observers' ability to analyse and report their conscious experience was impeached by John Watson (1913), who proposed a new methodological approach based on observable behaviour and performance. The criticism of introspective methods led eventually in laboratory psychology to the rejection of virtually all reports of thinking for several decades.

During the cognitive revolution in the 1950s and 1960s alternative types of verbal reports of thinking were used to gather information about cognitive structures and processes. Investigators soon discovered that some types of verbal reports were reactive, that is the performance of participants giving the reports differed from that of silent control subjects (Gagné and Smith, 1962). Other types of reports were directly inconsistent with the observed behaviour. In several reviews Herb Simon and I (Ericsson and Simon, 1980, 1993, 1998; Ericsson, 2002) showed that the detailed instructions and the methods to induce participants to give verbal reports influenced the validity and reactivity of collected verbal-report evidence. Most important, we proposed a new model for coordinating the processes responsible for generating 'think-aloud' reports on thinking with the ongoing task-directed processes to produce valid and non-reactive verbal reports. The same model explained how more complex verbalization processes induced changes in cognitive processes and led to reactive and invalid reports.

The Ericsson-Simon (1993) Model of Verbalization of Thinking has been accepted as a useful foundation for discussing introspection (see the entry on 'Psychology of Introspection' in the Routledge Encyclopedia of Philosophy by Von Eckardt (1998). This framework for collecting verbal reports on cognitive constructs, such as memory and rules, has even met the standards of evidence for behaviourists (Austin and Delaney, 1998). As a further recognition of its validity, protocol analysis now plays a central role in applied settings, such as in the design of surveys and interviews (Sudman, Bradburn and Schwarz, 1996) and user testing of computer tools and applications (Henderson *et al.*, 1995). It is fair to say that there has been a dramatic advance in the development of rigorous methods for collecting evidence on mediating cognitive processes and structures. At the same time, some investigators (Jack and Shallice, 2001) have viewed the Ericsson-Simon model as too constraining to study consciousness. Jack and Roepstorff (2002) propose a revision of the type of introspective methods embraced by pioneers such as William James (1890, p. 185), who recommended that we should 'first and foremost and always' be 'looking into our minds'.

I will argue here that a reintroduction of introspective methods must be based on a deep understanding of the past controversies over introspective reports. Now, a century later, we have a better understanding of some fundamental issues with evidence collected using the introspective method. In a brief historical overview I will describe how philosophers and pioneering psychologists discovered the problems associated with trusting observers' reports from their analysis of complex and dynamic mental states. I will also describe how they changed their research methods to avoid these problems and designed tasks to elicit reproducible performance. In the main section of this paper, I will describe how Herb Simon and I tackled the problems of verbal reports of thinking, specifically how we proposed theoretical accounts of the act of verbalizing thoughts with minimal interference and how we designed reporting procedures that yielded non-reactive overt expression of task-directed thinking. I will then review evidence on validity where this type of verbalization of thoughts is shown to be consistent with task analyses of how the task can be performed, with other types of process data, such as latencies and eye-fixations, and with experimental tests of proposed cognitive mechanisms. Herb Simon's and my goal was to design a highly constrained measurement situation where trust is not an issue. In performing these tasks not even devious participants would be able to produce verbalizations of thoughts that are inconsistent with their task-directed processes that mediate performance. At least, they would not be able to do so without violating our checks for validity.

Turning to my central argument about introspection I will show that verbal-report methods that attempt to collect more detailed observations on the mental states than can be unobtrusively verbalized are associated with reactive effects on performance. In my conclusions I will argue that such reporting methods can integrated into Herb Simon's and my theoretical framework by proposing detailed accounts of the additional cognitive processes that mediate the supplementary observations. Finally, I will propose how phenomena discovered by introspective methods can be captured through the design of new tasks that explicitly target those phenomena and where mediating processes can be studied with valid and non-reactive reports of task-directed thinking.

Varieties of Introspection and the Central Problems Encountered at the Beginning of the Twentieth Century

The controversy over 'imageless thought' should be seen as a culmination of efforts to refine introspective methods for studying thinking. The method of introspection changed from an informal method to study one's own experiences in the seventeenth and eighteenth centuries to a scientific method applied to specific mental phenomena in the laboratory in the late nineteenth century. To convey my view of the central problems of the introspective method, I will first describe my understanding of its historical development, its demise and its subsequent transformation into alternative types of verbal reports.

Historic sketch of the development of introspection

The observation of one's own spontaneous thinking by 'looking within' has a long history that can be traced back at least to the Greeks. Aristotle is generally given credit for the first systematic attempt to record and analyse the structure of thinking. He focused particularly on recalling a specific piece of information from memory. On one occasion he even reported on a specific sequence of thoughts corresponding to the recall of 'autumn': 'from milk, to white, from white to air, and this to fluid, from which one remembers autumn, the season one is seeking' (Aristotle translated by Sorabji, 1972, p. 56). More generally, Aristotle argued that thinking corresponds to a sequence of thoughts (see the boxes in Figure 1), where the brief transition periods between consecutive thoughts (see the arrows in Figure 1) do not contain any reportable information. Hence the processes that determined how one thought triggered the next thought could not be directly observed, but had to be inferred by retrospective reflection of the relations between consecutive thoughts. By examining his memory of which thoughts tended to follow one another, Aristotle inferred that previously experienced associations were the primary determining factor.

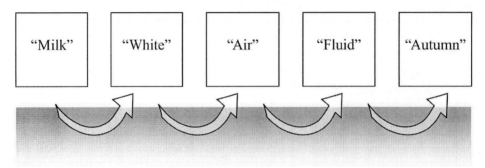

Figure 1. An illustration of Aristotle's view of thinking as a sequence of states with reportable thoughts, while the processing responsible for transitions between consecutive thoughts remains inaccessible.

Aristotle's account of thinking as a reportable sequence of thoughts has never been seriously challenged. This account also by implication recognizes limits for information that can be gained directly by introspection on thinking. However, philosophers in the seventeenth, eighteenth and nineteenth centuries (Ericsson and Crutcher, 1991) raised deeper questions about the nature of thoughts that could not be answered by such a simple description of thinking. For example, could all thoughts be described as mixtures of sensory images derived from past experiences? In order to evaluate these claims empirically, philosophers would typically relax and daydream, allowing their thoughts to wander. Once a thought emerged, it would be inspected carefully and studied to assess its sensory modality and components. The British philosopher David Hume even argued that these thoughts were as detailed as original perceptions (e.g. we cannot imagine a printed page of text without imaging every letter on the page at the same time) (James, 1884).

Towards the end of the nineteenth century, some philosophers proposed 'self experiments' to assess the accuracy and validity of such memory images. For example, Hamilton (1859) proposed that one could throw a number of pebbles behind one's back and then rapidly turn around and take a peek before closing one's eyes. When a memory image of such a glance was compared to the actual perceptual scene by opening one's eyes again, it was clear that the memory image contained accurate information about only a small number of specific pebbles. More generally, it was becoming clear that thoughts and mental images differed fundamentally from the perceptions and sensations corresponding to external objects.

External objects can be inspected and details about them can be noticed without changing their content and structure. In contrast, when thoughts and images are inspected the associated mental image also changes. For example, when one generates a visual image of rose and then focuses on one of its components, such as the petals, other aspects of the image of the rose are reported to change as well. The most influential laboratory psychologists of the late nineteenth century, Wilhelm Wundt (1897), argued that introspective analysis of experience was possible only when simple physical stimuli, such as points of light and brief sounds, were presented repeatedly to observers under controlled conditions. In fact, he argued forcefully that introspective analysis of complex thoughts as they emerge during spontaneous thinking was not possible, because such analysis would change the corresponding mental states and thus disrupt the course of thinking.

Around the beginning of the twentieth century investigators at the University of Würzburg generated considerable controversy with their attempts to study spontaneous thinking with introspective methods. Trained introspective observers were invited to their laboratory and asked prepared questions, such as 'Do you understand the following proverb, "We depreciate everything that can be explained"?' (Bühler, 1907). The observers were asked to give their answers to the questions as quickly as possible and then after each answered question they gave detailed retrospective reports about their thoughts that mediated the question-answering process. The retrospective reports were extensive and detailed. Most reported thoughts consisted of visual and auditory images, but some observers claimed to have experienced thoughts without any corresponding imagery (imageless thoughts).

The proposed existence of imageless thoughts had far-reaching theoretical implications for Wundt's theory that all 'experience' corresponds to neural activity. The original paper by Karl Bühler (1907) led to a heated exchange between him and Wilhelm Wundt (1907) who argued that these reports were artifacts of the reporting methods and the poor training of the observers. Ultimately, the most devastating aspect of this controversy was that it revealed that the issue of imageless thoughts could not be resolved empirically, which in turn impeached introspection as a scientific method by which to study thinking.

The central problems of introspective analysis of thoughts

The central problem of introspection was not so much 'looking within', but rather the problem of inducing the same mental states in many observers where the states were sufficiently stable to allow consistent judgments across observers. In fact, the introspective analysis of sensory stimuli in psychophysical studies of perception was never rejected by John Watson (1913), and this work continued uninterrupted for the whole twentieth century. In psychophysical studies experimenters could repeatedly present each of a collection of simple sensory stimuli, and the ability to discriminate these stimuli could be evaluated objectively. But no comparable method was available to reproduce the same thought or complex mental state for many different participants. Therefore, it was necessary to trust the observers' reports of their idiosyncratic thoughts, because it was not possible to assess the observers' thoughts or mental states independently in order to allow objective verification of the accuracy of their reports.

(a) From trust to reproducible performance. At the beginning of the twentieth century researchers were naive about problems of trust. For example, in studies of memory the participants were presented with stimuli and later the same stimuli were presented to assess memory and provide introspective reports. The subjects performing the introspections knew that the same stimuli were shown both at presentation and at the recognition test, so one would have to trust their reports that they remembered the stimuli. New procedures for objective tests of memory were later developed; the participants were given recognition tests where they had to identify the presented stimuli among a mixture of old and not-previously-seen stimuli. This measurement procedure relieved the need to trust the participants' reports of memory, because highly accurate recognition (reliably higher than guessing) provides proof that the participants remembered the presented stimuli. Even more compelling is the case when a participant could recall most of the presented stimuli. The probability is small that a participant can correctly guess which word was presented from items sampled from a pool of several thousand words.

Subsequent research relied on these objective tests of memory to evaluate the validity of introspective judgments of participants. For example, a large body of research has attempted to relate the level of accurate recall of a presented picture to the reported vividness of the memory (McKelvie, 1995; Richardson, 1988). To everyone's surprise, no clear relation between the amount of accurately recalled information and reported vividness has been found. Participants who reported recalling a presented stimulus as vividly and clearly as if it remained visible did not recall more accurate information than those who reported diffuse memory images. These and other puzzling findings, such as the reported persistence of visual eidetic images (Haber, 1979), confirmed the opinions of many experimental psychologists that introspective judgments about experience were frequently misleading and inconsistent with measures of performance.

More generally, experimental psychologists developed standardized tests with stimuli and instructions where the same pattern of performance could be replicated in the same laboratory as well as different ones. Furthermore, psychologists redirected their research away from complex mental processes, such as thinking, and towards processes that were unaffected by prior experience and knowledge. For example, subjects were given well-defined simple tasks, such as memorization of lists of nonsense syllables, e.g. XOK, ZUT. In these tasks it is easy to measure objective performance and thus there is no need to trust participants' honesty or their willingness to cooperate. In addition, the formation of simple basic associations in memory led experimenters to predict no cognitive mediation and thus no reports of mediating thoughts, and the issue of trusting subjects' verbal reports appeared to become essentially irrelevant.

(b) Studying complex thought with non-introspective verbal reports. The interest in studying complex processes of thinking did not completely stop. Interestingly, the first investigator to come up with a new method for studying thinking was John Watson (1920). He instructed a friend to solve a problem and asked him to 'think aloud' while working on it. According to Watson, thinking was accompanied by covert neural activity of the speech apparatus that is 'inner speech'. Hence thinking aloud did not require observations by any hypothetical introspective capacity, and thus all that was necessary to think aloud was merely to give overt expression to these subvocal verbalizations. Many other investigators proposed similar types of instructions to give concurrent verbal expression of one's thoughts (see Ericsson and Simon, 1993, for a brief historical review).

Renewed Interest in Verbal Reports of Thinking During the Cognitive Revolution

The cognitive revolution in the 1950s and 1960s brought renewed interest in higher-level cognitive processes. Investigators started to explore how problem solving, concept formation and decision making could be explained by mediating thought processes. Cognitive theories were proposed in which strategies, concepts and rules were central to the account of human learning and problem solving (Miller, Galanter and Pribram, 1960). Information-processing theories (Newell and Simon, 1972) sought computational models that could regenerate human performance on well-defined tasks by the application of explicit procedures. Much of the evidence for these complex mechanisms was derived from self-observation, informal interviews and systematic questioning of participants.

Almost immediately some investigators raised concerns about the validity of these data (Ericsson and Simon, 1993). For example, Robert Gagné and his colleagues (Gagné and Smith, 1962) demonstrated that requiring participants to verbalize reasons for each move in the Tower of Hanoi reduced the number of moves in the solutions and improved transfer to more difficult problems when compared to a silent control condition. Other investigators criticized the validity and accuracy of the retrospective verbal reports. For example, Nisbett and Wilson (1977) showed that participants frequently gave explanations that were inconsistent with their observed behaviour. Consequently, it might appear at first glance that all of these types of verbal reporting would have to be rejected on the same grounds that led to the rejection of introspection. Herb Simon and I (Ericsson and Simon, 1980; 1993), however, were able to show that the methodology had advanced and that it was possible to identify conditions where participants are able to produce consistently valid non-reactive reports of their thinking.

Towards Valid and Non-Reactive Verbal Reports of Thinking

The most significant advances since the earlier controversy over 'imageless thought' concerned the methodology for inducing thinking by presenting well-defined tasks as well as the formal theoretical analysis of possible mediating processes (task analysis).

Task analysis

The introspectionists in Würzburg presented their observers with tasks to induce spontaneous thinking. Their tasks, such as whether the observers had understood a presented proverb, didn't even have a correct answer. A devious participant could conceivably decide to answer 'yes' or 'no' even before he heard the proverb, and the experimenter would not be able to tell. Consequently, cognitive researchers developed collections of tasks that would induce some specific mental activity and where participants could emit consistently accurate responses only after seeing the stimulus and after some cognitive processing. For example, it would be impossible to guess the correct answers before seeing a maths problem, such as '258 + 893 = ?' or '24 × 36 = ?'.

Completing a series of tasks correctly requires relevant knowledge. In fact it is possible for the investigator to identify various procedures that people could possibly use, in light of their prior knowledge and skills, to generate correct answers. This type of analysis (task analysis) provides a set of possible thought sequences for the successful performance of a task, where the application of each alternative procedure is associated with a different sequence of thoughts (intermediate steps). Let me illustrate how this general type of analysis can be applied to the mentally demanding task of multiplying two 2-digit numbers 'in one's head'. Typically, many adults have acquired basic mathematical knowledge: they know their multiplication table and only the standard 'pencil and paper' procedure taught in school for solving multiplication problems. Accordingly, one can predict that most adults will solve a problem such as 36 × 24 'in their head' by first calculating 4 × 36 then adding 20 × 36. But this specific problem can be solved using alternative methods that are more efficient for adults who know some of the squares of 2-digit numbers. People with more advanced knowledge of mathematics may recognize that 24 × 36 is equivalent to (30 – 6) × (30 + 6) and use the formula $(a + b) \times (a - b) = a^2 - b^2$, thus calculating 36 × 24 as $30^2 - 6^2 = 900 - 36 = 864$. Other subjects may recognize other shortcuts, such as $36 \times 24 = (3 \times 12) \times (2 \times 12) = 6 \times 12^2 = 6 \times 144 = 864$.

Observations on processes that mediate successful task performance

Whereas introspection requires additional observation and analysis by the participant, cognitive researchers obtain information on concurrent processes by simply observing the participants completing the task. Although a covert sequence of thoughts generated during the performance of a task (illustrated in the centre of Figure 2) is never directly observable, it is associated with several types of

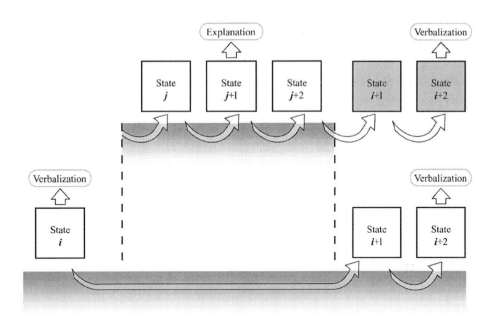

Figure 2. An illustration of a covert sequence of thoughts (centre) generated in response to a presented task along with its associated observable non-verbal indicators.

observable indicators. At the top of Figure 2, the total time required by subjects to generate their answer (response latency) can be viewed as the sum of the component times for generation of each thought, such as intermediate products and results, when the corresponding thought processes are not overlapping. From the task analysis it is often possible to infer the number of intermediate steps that would be required for solving different problems with alternative methods. From an analysis of average latencies across a set of different problems it will be possible to determine which procedures can best account for the subjects' observed latency pattern. In addition, most tasks are presented to subjects visually and it is necessary for subjects to direct their visual gaze sequentially to process the essential presented information, while an eye-tracking device can register their sequences of eye-fixations. Given that different hypothesized procedures typically predict different sequences of fixating displayed information, it is possible to identify the procedure that shows the highest agreement with the observed eye-fixations.

Other types of observations on thought processes can be collected by instructing participants to provide observable evidence on their thought processes. First, the subjects can be instructed to verbalize their thinking concurrently with performing the task (see Figure 2). Alternatively, the subjects can be asked to give a retrospective report of their thought sequence immediately after the completion of the task. Finally, the subject can give a general description of her strategies (post-session reports) once she has completed numerous tasks during the testing session. All of these types of verbal reports share the problem of introspection however, their generation can be explained only by proposing some additional cognitive activity because participants do not spontaneously generate these reports when they perform tasks.

Models of the generation of verbal reports

On the basis of their theoretical analysis, Ericsson and Simon (1993) argued that the closest connection between thinking and verbal reports is found when subjects verbalize thoughts generated during task completion (see Figure 2). When subjects are asked to think aloud, some of their verbalizations correspond to merely vocalizing 'inner speech', which would otherwise have remained inaudible. Non-verbal thoughts can also be given verbal expression by brief expressions, labels and referents. For example, when one subject was asked to think aloud while mentally multiplying 36×24 on two test occasions one week apart the following protocols were obtained:

1. OK, 36 times 24, um, 4 times 6 is 24, 4, carry the 2, 4 times 3 is 12, 14, 144, 0, 2 times 6 is 12, 2, carry the 1, 2 times 3 is 6, 7, 720, 720, 144 plus 720, so it would be 4, 6, 864.
2. 36 times 24, 4, carry the — no wait, 4, carry the 2, 14, 144, 0, 36 times 2 is, 12, 6, 72, 720 plus 144, 4, uh, uh, 6, 8, uh, 864.

In these two examples, the reported thoughts are not introspectively analysed into their perceptual or imagery components, but merely verbally expressed and referenced, such as 'carry the 1', '36' and '144 plus 720'. Similarly, subjects are not asked to describe or explain how they solve these problems. Instead, they are instructed to remain focused on solving the problem and merely to give verbal expression to those thoughts that emerge in attention while generating the solution under normal (silent) conditions.

If the act of verbalizing subjects' thought processes doesn't change the sequence of thoughts while completing well-defined tasks, then subjects' task performance should not change as a result of thinking aloud. In a comprehensive review of more than forty studies, Ericsson and Simon (1993) found no evidence that merely giving concurrent verbal expression to one's thoughts (cf Ericsson and Simon's procedure for inducing 'think aloud') altered accuracy of performance compared to that of subjects who completed the same tasks silently under otherwise similar conditions. But some studies showed that participants would take somewhat longer to complete the tasks while thinking aloud — presumably because of the additional time required for completing the overt vocalization of the verbal expression of the thoughts.

The same theoretical framework can also explain how other types of concurrent verbal report procedures would change cognitive processes. For example, when subjects are instructed to explain or carefully describe their thoughts, in contrast to merely verbalizing each thought as it emerges, they are not able to remain completely focused on the task. To be able to verbalize the required explanations and descriptions the participants need to change their thought processes to generate the corresponding thoughts, as is illustrated in Figure 3. This

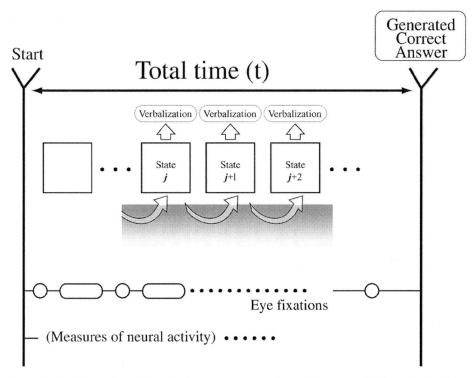

Figure 3. An illustration of how the interspersed cognitive activity of explaining and describing can change the sequence of thoughts as thinking is resumed.

additional cognitive activity must necessarily change the sequence of mediating thoughts. Therefore when participants resume their task-directed processes after the explanation, the sequence of thoughts will differ and thus lead to changes in the accuracy of performance. Our reviews (Ericsson and Simon, 1993, 1998; Ericsson, 2002) have found that instructions to explain one's thinking or to give extended detailed descriptions of presented stimuli (cf. 'verbal overshadowing' reviewed by Schooler, Fiore and Brandimonte, 1997) are associated with reliable changes in memory and task performance.

In sum, after brief instruction and familiarization in giving verbal reports, subjects can think aloud without any systematic changes to the sequential structure of their thought processes (see Ericsson and Simon 1993, for detailed instructions and associated warm-up tasks recommended for laboratory research). The fact that subjects must already possess the necessary skills for efficient verbalization of thoughts is consistent with the extensive evidence on the acquisition of self-regulatory private speech during childhood (Diaz and Berk, 1992) and on the spontaneous vocalization of inner speech by adults, especially in noisy environments (Ericsson and Simon, 1993).

Reproducibility of cognitive phenomena, performance and associated process data

In sharp contrast to the tasks used in traditional research on introspection, cognitive psychologists carefully design well-structured experimental tasks where the successful generation of accurate responses is mediated by the cognitive activity under investigation. For example, it is possible to design tasks that predictably induce one or more of the conscious processes (Type-C processes, such as noticing, planning and monitoring) that Jack and Shallice (2001) propose mediate introspective judgments.

The research methodology attempts to push participants to reach their best consistent performance for the collection of tasks by providing opportunities for 'warm-up' and familiarization with the task before collecting the experimental data on performance (Ericsson and Oliver, 1988). Under these constrained performance conditions it should not even be possible for individuals to exhibit a performance that would be superior to the one observed. Similarly it should not be possible for devious participants to 'fake' the observations on the processes mediating their task performance, that is to intentionally produce eye-movements or verbal reports consistent with one method while proficiently generating the answer to the tasks with a different method.

The processes mediating performance are hardly ever perfectly reproduced for the same task, even when the same subject is tested (cf. the 'think aloud' protocols above, collected a week apart). Even in the case of thinking aloud, where the connection between thoughts and reports is the closest, there is no perfect mapping. This lack of one-to-one correspondence is primarily due to the fact that thoughts that pass through attention are not always verbalized. Most important, verbal reports do not differ in kind from other observations of processes, such as latencies and eye movements, and verbal expression of thoughts is influenced by many uncontrollable factors and thus will vary to some degree from trial to trial.

Validity of concurrent verbal reports and other process data

When we restrict the analysis to those thoughts that are verbalized, the evidence for validity is consistently strong (Ericsson and Simon, 1993). First, when the participants complete the task correctly, their verbalized information is in agreement with one of the sequences of thoughts derived from the task analysis, which had been completed before the observation of the verbal reports. For example, the sample protocols on mental multiplication (reported above) are consistent with only one of the methods uncovered in the task analysis — namely the 'paper-and-pencil' method. Even if the verbalized information had been more sparse for a highly skilled subject and only contained '144' and '720', the reported information would still have been sufficient to reject each of the alternative multiplication methods, because neither of those methods involves generating both of the reported intermediate products. The general finding that a task analysis can identify, a priori, the specific intermediate products that are later verbalized by subjects during their problem solutions, provides compelling evidence that the concurrent verbalizations reflect the processes that mediate the actual generation of the correct answer.

Second, the verbal reports are only one of several different kinds of observable indicators of the same thought process (see Figure 2). Given that each kind of empirical indicator can be separately recorded and analysed, it is possible to evaluate the agreement of different types of data. Ericsson and Simon (1993) found that the solution methods derived from an analysis of the verbal reports were consistent with those derived from analyses of response latencies and sequences of eye-fixations.

Third, the hypothesized mechanisms mediating thinking and the observed verbalizations can be examined with experimental methods (Ericsson and Simon, 1993). In particular, the reproducibly superior performance of expert and skilled performers offers a unique opportunity for studies of complex cognitive mechanisms using verbal reports. When the sequence of concurrently verbalized thoughts during the execution of representative tasks is reproducible for single experts, then it is possible to experimentally test the validity of the hypothesized cognitive mechanisms by traditional experimental methods. For example, verbal reports of people with exceptional memory have then been evaluated experimentally by presenting them with altered memory tasks where their performance has been predictably reduced in a decisive manner (Ericsson, 1988; Ericsson and Kintsch, 1995; Ericsson, Patel and Kintsch, 2000).

In sum, think-aloud reports provide the most informative data available on thinking during cognitive tasks. Those aspects of verbal reports that can be validated against other sources suggest a close correspondence between the reports and the cognitive processes that produce intermediate results in attention. On the other hand, the reports are not infallible; subjects may occasionally make speech errors and omit articulation of disambiguating information in their verbalizations of thoughts.

Other Types of Verbal Reports of Thinking and Cognition

The main goal for Herb Simon and me (Ericsson and Simon, 1980, 1993) was to identify the conditions and the verbal-report procedure that would elicit valid and verifiable verbalizations of thinking. It is, however, possible to use our model for these verbalization processes of thoughts during performance in discussing other types of verbal reports.

The type of retrospective verbal report that is most closely related to 'think aloud' verbalization involves asking the participants to report their thoughts immediately after the completion of a given trial. When the time to generate the response is brief (1s to 5s), it is likely that the participants can recall their sequence of thoughts reasonably accurately. Our review (Ericsson and Simon, 1993) showed that when participants are asked merely to recall their thoughts, the reported information is consistent with other observations of the same processes, such as latencies.

In contrast to this type of retrospective report, most other verbal reports of thinking require cognitive activities that go beyond immediate recall sequences of already generated thoughts. Ericsson and Simon (1980, 1993) identified two additional cognitive activities that contribute to the decreased validity of the verbally reported information. The first arises when the investigators try to obtain more information than the subjects' thought sequences can provide. For example, some investigators ask subjects *why* they responded in a certain way. Sometimes the subjects' recalled thoughts may provide sufficient information for an answer, but typically the subjects need to go beyond any retrievable memory of their processes to give an answer. As subjects can access the end-products of their cognitive processes only during perception and memory retrieval, they cannot report why only one of several logically possible thoughts entered their attention and thus must speculate to generate answers to such questions. In support of this argument Nisbett and Wilson (1977) found that subjects' responses to why-questions in many circumstances were as inaccurate as those given by other subjects that merely observed and explained another subject's behaviour. Since our review in 1993, Schooler, Fiore and Brandimonte (1997) have presented similar evidence for invalidity of retrospective reports when participants were instructed to give extended detailed descriptions of them (cf. 'verbal overshadowing'). In an attempt to replicate these effects, Meissner, Brigham and Kelley (2001) found no effect of merely recalling what the subjects could confidently remember (cf. the standard retrospective report). They were also able to pinpoint the source of invalidity to aspects of verbal-overshadowing instructions where participants were encouraged to keep producing verbal descriptions, even if they started to feel that they were guessing. In a review of a large number of related studies I (Ericsson, 2002) found no effects of merely giving verbal expression to memories, but consistent effects of attempting to report more than subjects could confidently recall.

Second, investigators often ask subjects to describe their methods for solving problems at the end of the experiment, when they have completed a long series of different tasks. If subjects generated and consistently applied a general strategy for solving all of the problems, they should be able to respond to such requests easily with a single memory retrieval. But subjects typically employ many methods and shortcuts and even change their strategies during the experiment, through learning. Under such circumstances subjects would have great difficulty describing a *single* strategy used consistently throughout the experiment, even in the unlikely event that they were motivated and able to recall most of the relevant thought sequences. It is therefore not surprising that subjects' descriptions are imperfectly related to their averaged performance during the entire experiment.

Conclusion

There is compelling evidence that some types of verbal-report procedures lead to reactive effects and elicit verbal reports that are inconsistent with the participants' observed performance (Ericsson and Simon, 1993). It is therefore not reasonable to ask the scientific community to trust verbal reports in general. Grouping all types of verbal-report procedures together will bring us back to the unproductive controversies of the beginning of the twentieth century and to the renewed criticisms of verbal reports during the 1960s and 1970s. To avoid an indiscriminate rejection of verbal reports as data, it is essential that we distinguish different types of verbal reports. Herb Simon and I (Ericsson and Simon, 1980, 1993) identified a theoretically motivated procedure to elicit valid and non-reactive verbalization of thoughts. We proposed the existence of processes that could vocalize and verbalize thoughts passing through attention without disrupting the flow of thinking. Furthermore, we (Ericsson and Simon, 1980, 1993) showed in comprehensive reviews of a large number of empirical studies that our recommended procedures for thinking aloud and talking aloud provided valid verbalizations of thoughts where the generation did not influence the sequence of thoughts. In contrast, procedures demanding explanations and detailed descriptions led to reactive effects and thus would provide an inaccurate account of thought under typical (silent) task conditions.

In our book (Ericsson and Simon, 1993) we invited researchers to extend our theoretical analysis and methods of validation to other types of verbal-reporting procedures. Our recommended procedures for eliciting think aloud and talk aloud may well be unduly constraining, as Jack and Shallice (2001) propose. It is entirely possible that alternative and more introspective procedures would provide more information. The critical question is *how* the proposed additional introspective processes are coordinated with the task-directed processes to yield verbal reports of extra information beyond that included in think aloud and immediate retrospective reports. If the introspective processes are additional to the regular task-directed processes, wouldn't it be possible, or even likely, that these processes might be reactive and alter performance? In this paper I have reviewed empirical evidence for how requirements to go beyond the heeded sequences of thoughts, such as by generating explanations and extended detailed descriptions, results in reactivity and alters performance and memory. It would therefore be important for the proponents of new introspective procedures to account for these problems of validity and reactivity and describe if and how the new procedures can avoid those types of interference. If we follow an orderly approach for extending the repertoire of verbal-report procedures it would help us to steadily accumulate a body of valid evidence of verbal reports on thinking and, whenever empirically warranted, expand the repertoire of useful verbal-report procedures.

In our book (Ericsson and Simon, 1993), we argued that it is not necessary to dismiss reactive verbal report procedures. It is better to view the verbal reports as reports for a different task, where the demands for the verbalizations are an integral part. For example, most educators are very interested in understanding the reactive yet typically beneficial effects of instructing students to explain their performance. Recent research inducing students to generate self-explanations is a very interesting development in that direction (Chi *et al.*, 1994; Renkl, 1997). The requirement to generate self-explanations has been shown to have a reliably different effect on problem-solving performance than does merely thinking aloud (Neuman and Schwarz, 1998). Similarly, verbal-report instructions requiring explanations influence performance, whereas thinking aloud doesn't alter performance compared to a silent control condition (Berardi-Coletta et al., 1995). For anyone interested in effective education there is no single correct procedure for inducing verbalizations during problem solving. In fact, a detailed analysis of the different verbalizations elicited during 'think aloud' and 'explain' instructions should provide investigators with an effective tool to identify those induced cognitive processes that are associated with desired changes (improvements) in the subjects' task performance, memory and understanding.

One could make a similar argument that asking individuals to introspect while performing a task is reactive but merely changes the task demands. One needs to consider that introspecting while performing the task differs from the traditional task in some important ways. In the traditional task the participants are helped to find a stable performance that allows repeated completion of a series of similar tasks in a very efficient and reproducible manner. In contrast, the introspecting subjects are instructed to engage in additional observation and noticing. Which aspects of the cognitive processes and for how long these aspects are observed are likely to be unpredictable to both the subjects and the experimenter. As a consequence, the traditional methods of validation, such as task analysis, agreement between reports and performance characteristics (e.g. latencies and eye-fixations) and designed experimental tests, will not be available as tools for the analysis of open-ended introspective reports. This type of introspective report can still provide valuable opinions and ideas that might lead to the generation of interesting and more targeted hypotheses. One of the pioneers of the study of the brain, Lashley (1923, p. 352), said: 'introspection may make the preliminary survey, but it must be followed by the chain and transit of objective measurement'. Many cognitive phenomena, such as afterimages or feelings of knowing, were originally discovered by introspective observation, presumably while the person was engaged in some other task-related activity. Once this type of experience has been recorded and proposed to reflect a general phenomenon it is possible for investigators to try to design tasks where this specific phenomenon can be reliably induced in many subjects. For example, subjects can be presented with bright colour stimuli and then asked to name the corresponding colour in the afterimage (cf. Comstock and Kittredge, 1922). To study the 'feeling of knowing', subjects have been asked to estimate the probability that they will be able to recognize the correct name of the capitol of a presented name of a country, where the estimates can be evaluated against performance on subsequent memory tests (Gruneberg and Monks, 1974). Once suitable tasks with well-defined criteria for performance have been developed, the thought processes of associated phenomena can then be studied with non-reactive reporting methods, such as think aloud and immediate retrospective reports.

In the current exciting quest to better understand consciousness it is hard to overestimate the importance of rigorous data-collection methods that produce independent scientific evidence. Crucial future advances in cognitive science and cognitive psychology will, by necessity, require that we keep rejecting prevailing theories and fundamental intuitions. These types of advances will only be possible when we have methods that provide reproducible empirical evidence; only such evidence can compel scientists to change their beliefs and thus be capable of successfully resolving current and future theoretical controversies.

Acknowledgements

This article was prepared while the author was a Fellow at the Center for Advanced Study in the Behavioral Sciences, Stanford. I am grateful for the financial support provided by the John D. and Catherine T. MacArthur Foundation, Grant #32005–0. I also want to thank Kathleen Much and Natalie Sachs-Ericsson for very helpful comments on earlier drafts of this article.

References

Austin, J. and Delaney, P.F. (1998), 'Protocol analysis as a tool for behavior analysis', *Analysis of Verbal Behavior*, **15**, pp. 41–56.
Berardi-Coletta, B., Buyer, L.S., Dominowski, R.L. and Rellinger, E.R. (1995), 'Metacognition and problem solving: a process-oriented approach', *Journal of Experimental Psychology: Learning, Memory, & Cognition*, **21**, pp. 205–23.
Bühler, K. (1907), 'Tatsachen und Probleme zu einer Psychologie der Denkvorgänge: I. Uber Gedanken [Facts and problems in a psychology of thinking: I. On thoughts]', *Archiv für die gesamte Psychologie*, **9**, pp. 297–365.
Chi, M.T.H., de Leeuw, N., Chiu, M.-H. and LaVancher, C. (1994), 'Eliciting self-explanations improves understanding', *Cognitive Science*, **18**, pp. 439–77.
Comstock, C. and Kittredge, H. (1922), 'An Experimental Study of Children as Observers', *American Journal of Psychology*, **33**, pp. 161–77.
Diaz, R.M. and Berk L.E. (1992), *Private Speech: From Social Interaction to Self-Regulation* (Hillsdale, NJ: Erlbaum).
Ericsson, K.A. (1988), 'Analysis of memory performance in terms of memory skill', in *Advances in the Psychology of Human Intelligence*, Vol. 4, ed. R.J. Sternberg (Hillsdale, NJ: Erlbaum), pp. 137–79.
Ericsson, K.A. (2002), 'Toward a procedure for eliciting verbal expression of nonverbal experience without reactivity: interpreting the verbal overshadowing effect within the theoretical framework for protocol analysis', *Applied Cognitive Psychology*, **16**, pp. 981–7.
Ericsson, K.A. and Crutcher, R.J. (1991), 'Introspection and verbal reports on cognitive processes — two approaches to the study of thought processes: a response to Howe', *New Ideas in Psychology*, **9**, pp. 57–71.
Ericsson, K.A. and Kintsch, W. (1995), 'Long-term working memory', *Psychological Review*, **102**, pp. 211–45.
Ericsson, K.A. and Oliver, W. (1988), 'Methodology for laboratory research on thinking: task selection, collection of observation and data analysis', in *The Psychology Of Human Thought*, ed. R.J. Sternberg and E.E. Smith (Cambridge: Cambridge University Press), pp. 392–428.
Ericsson, K.A., Patel, V.L. and Kintsch, W. (2000), 'How experts' adaptations to representative task demands account for the expertise effect in memory recall: comment on Vicente and Wang (1998)', *Psychological Review*, **107**, pp. 578–92.
Ericsson, K.A. and Simon, H.A. (1980), 'Verbal reports as data', *Psychological Review*, **87**, pp. 215–51.
Ericsson, K.A. and Simon, H.A. (1993), *Protocol Analysis; Verbal Reports As Data* (revised edition; original edition published 1984) (Cambridge, MA: Bradford Books/MIT Press).
Ericsson, K.A. and Simon, H.A. (1998), 'How to study thinking in everyday life: contrasting think-aloud protocols with descriptions and explanations of thinking', *Mind, Culture, & Activity*, **5**, pp. 178–86.

Gagné, R.H. and Smith, E.C. (1962), 'A study of the effects of verbalization on problem solving', *Journal of Experimental Psychology*, **63**, pp. 12–18.
Gruneberg, M.M. and Monks, J. (1974), ' "Feeling of knowing" and cued recall', *Acta Psychologica*, **38**, pp. 257–65.
Haber, R.N. (1979), 'Twenty years of haunting eidetic imagery: where's the ghost?', *Behavioral & Brain Sciences*, **2**, pp. 583–629.
Hamilton, Sir W. (1859), *Lectures on Metaphysics and Logic*, Vol. I (Edinburgh: Blackwood).
Henderson, R.D., Smith, M.C., Podd, J. and Varela-Alvarez, H. (1995), 'A comparison of the four prominent user-based methods for evaluating the usability of computer software', *Ergonomics*, **39**, pp. 2030–44.
Jack, A.I. and Roepstorff, A. (2002), 'Introspection and cognitive brain mapping: from stimulus-response to script-report', *Trends in Cognitive Sciences*, **6**, pp. 333–9.
Jack, A.I. and Shallice, T. (2001), 'Introspective physicalism as an approach to the science of consciousness', *Cognition*, **79**, pp. 161–96.
James, W. (1884), 'On some omissions of introspective psychology', *Mind*, **9**, pp. 1–26.
James, W. (1890), *Principles of Psychology* (New York: Holt).
Lashley, K.S. (1923), 'The behavioristic interpretation of consciousness II', *Psychological Review*, **30**, pp. 329–53.
McKelvie, S.J. (1995), 'The VVIQ and beyond: vividness and its measurement', *Journal of Mental Imagery*, **19**, pp. 197–252.
Meissner, C.A., Brigham, J.C. and Kelley, C.M. (2001), 'The influence of retrieval processes in verbal overshadowing', *Memory and Cognition*, **29**, pp. 176–86.
Miller, G.A., Galanter, E. and Pribram, K.H. (1960), *Plans and the Structure of Behavior* (New York: Holt, Rinehart, and Winston).
Neuman, Y. and Schwarz, B. (1998), 'Is self-explanation while solving problems helpful? The case of analogical problem-solving', *British Journal of Educational Psychology*, **68**, pp. 15–24.
Newell, A. and Simon, H.A. (1972), *Human Problem Solving* (Englewood Cliffs, NJ: Prentice-Hall).
Nisbett, R.E. and Wilson, T.D. (1977), 'Telling more than we can know: verbal reports on mental processes', *Psychological Review*, **84**, pp. 231–59.
Renkl, A. (1997), 'Learning from worked-out examples: a study on individual differences', *Cognitive Science*, **21**, pp. 1–29.
Richardson, J.T.E. (1988), 'Vividness and unvividness: reliability, consistency, and validity of subjective imagery ratings', *Journal of Mental Imagery*, **12**, pp. 115–22.
Schooler, J.W., Fiore, S.M. and Brandimonte, M.A. (1997), 'At a loss for words: verbal overshadowing of perceptual memories', *The Psychology of Learning and Motivation*, **37**, pp. 291–340.
Sorabji, R. (1972), *Aristotle on Memory* (Providence, RI: Brown University Press).
Sudman, S., Bradburn, N.M. and Schwarz, N. (ed. 1996), *Thinking About Answers: The Application of Cognitive Processes to Survey Methodology* (San Francisco, CA: Jossey-Bass).
Von Eckardt, B. (1998), 'Psychology of Introspection', in *Routledge Encyclopedia of Philosophy*, ed. E. Craig (London: Routledge), pp. 842–6.
Watson, J.B. (1913), 'Psychology as the behaviorist views it', *Psychological Review*, **20**, pp. 158–77.
Watson, J.B. (1920), 'Is thinking merely the action of language mechanisms?', *British Journal of Psychology*, **11**, pp. 87–104.
Wundt, W. (1897), *Outlines of Psychology*, trans. C.H. Judd (Leipzig: Wilhelm Engelmann).
Wundt, W. (1907), 'Uber Ausfrageexperimente und über die Metoden zur Psychologie des Denkens [On interrogation experiments and on the methods of psychology of thinking]', *Philosophische Studien*, **3**, pp. 301–60.

Daniel C. Dennett

Who's On First?

Heterophenomenology Explained

There is a pattern of miscommunication bedeviling the people working on consciousness that is reminiscent of the classic Abbott and Costello 'Who's on First?' routine. With the best of intentions, people are talking past each other, seeing major disagreements when there are only terminological or tactical preferences — or even just matters of emphasis — that divide the sides. Since some substantive differences also lurk in this confusion, it is well worth trying to sort out. Much of the problem seems to have been caused by some misdirection in my apologia for *heterophenomenology* (Dennett, 1982; 1991), advertised as an explicitly *third*-person approach to human consciousness, so I will try to make amends by first removing those misleading signposts and sending us back to the real issues.

On the face of it, the study of human consciousness involves phenomena that seem to occupy something rather like another dimension: the private, subjective, '*first*-person' dimension. Everybody agrees that this is where we start. What, then, is the relation between the standard 'third-person' objective methodologies for studying meteors or magnets (or human metabolism or bone density), and the methodologies for studying human consciousness? Can the standard methods be extended in such a way as to do justice to the phenomena of human consciousness? Or do we have to find some quite radical or revolutionary alternative science? I have defended the hypothesis that there is a straightforward, conservative extension of objective science that handsomely covers the ground — *all* the ground — of human consciousness, doing justice to all the data without ever having to abandon the rules and constraints of the experimental method that have worked so well in the rest of science. This third-person methodology, dubbed heterophenomenology (phenomenology *of another* not oneself), is, I have claimed, the sound way to take the *first* person point of view as seriously as it can be taken.

To place heterophenomenology in context, consider the following ascending scale of methods of scientific investigation:

experiments conducted on anaesthetized animals;
experiments conducted on awake animals;
experiments on human subjects conducted in 'behaviorese'
— subjects are treated as much as possible like laboratory rats, trained to criterion with the use of small rewards, with minimal briefing and debriefing, etc.;
experiments in which human subjects collaborate with experimenters
— making suggestions, interacting verbally, telling what it is like.

Only the last of these methods holds out much hope of taking human subjectivity seriously, and at first blush it may seem to be a first-person (or, with its emphasis on communicative interaction with the subjects, second-person) methodology, but in fact it is *still* a third-person methodology if conducted properly. It is heterophenomenology.

Most of the method is so obvious and uncontroversial that some scientists are baffled that I would even call it a method: basically, you have to take the vocal sounds emanating from the subjects' mouths (and your own mouth) and *interpret* them! Well of course. What else could you do? Those sounds aren't just belches and moans; they're speech acts, reporting, questioning, correcting, requesting, and so forth. Using such standard speech acts, other events such as button-presses can be set up to be interpreted as speech acts as well, with highly specific meanings and fine temporal resolution. What this interpersonal communication enables you, the investigator, to do is to compose a catalogue of *what the subject believes to be true about his or her conscious experience.* This catalogue of beliefs fleshes out the subject's *heterophenomenological world*, the world according to S — the subjective world of one subject — not to be confused with the real world. The total set of details of heterophenomenology, plus all the data we can gather about concurrent events in the brains of subjects and in the surrounding environment, comprise the total data set for a theory of human consciousness. It leaves out no objective phenomena and no subjective phenomena of consciousness.

Just what kinds of things does this methodology commit us to? Beyond the unproblematic things all of science is committed to (neurons and electrons, clocks and microscopes, ...) just to *beliefs* — the beliefs expressed by subjects and deemed constitutive of their subjectivity. And what kind of things are beliefs? Are they sentences in the head written in brain writing? Are they non-physical states of dualist ectoplasm? Are they structures composed of proteins or neural assemblies or electrical fields? We may stay maximally noncommittal about this by adopting, at least for the time being (I recommend: for ever), the position I have defended (Dennett, 1971; 1987; 1991) that treats beliefs from *the intentional stance* as *theorists' fictions* similar to centres of mass, the Equator, and parallelograms of forces. In short, we may treat beliefs as *abstractions* that measure or describe the complex cognitive state of a subject rather the way horsepower indirectly but accurately measures the power of engines (don't look in the engine for the horses). As Churchland (1979) has pointed out, physics already has hundreds of well-understood measure predicates, such as *x has weight-in-grams n*, or *x is moving up at n meters per second*, which describe a physical property of x by relating it to a *number*. Statements that attribute beliefs using the standard *propositional attitude* format, *x believes that p*, describe x's internal state by relating it to a *proposition*, another kind of useful abstraction, systematized in logic, not arithmetic. We need beliefs anyway for the rest of social science, which is almost entirely conducted in terms of the intentional stance, so this is a conservative exploitation of already quite well-behaved and well-understood methods.

A catalogue of beliefs about experience is not the same as a catalogue of experiences themselves, and it has been objected (Levine, 1994) that 'conscious experiences themselves, not merely our verbal judgments about them, are the primary data to which a theory must answer'. But how, in advance of theory, could we catalogue the experiences themselves? We can see the problem most clearly in terms of a nesting of proximal sources that are presupposed as we work our way up from raw data to heterophenomenological worlds:

(a) 'conscious experiences themselves'
(b) beliefs about these experiences
(c) 'verbal judgments' expressing those beliefs
(d) utterances of one sort or another

What are the 'primary data'? For heterophenomenologists, the *primary* data are the utterances, the *raw,* uninterpreted data. But before we get to theory, we can interpret these data, carrying us via (c) speech acts to (b) beliefs about experiences.[1] These are the primary *interpreted* data, the pretheoretical data, the *quod erat explicatum* (as organized into heterophenomenological worlds), for a science of consciousness. In the quest for primary data, Levine wants to go all the way to (a) conscious experiences themselves, instead of stopping with (b) subjects' beliefs about their experiences, but this is not a good idea. If (a) outruns (b) — if you have conscious experiences you don't believe you have — those extra conscious experiences are just as inaccessible *to you* as to the external observers. So Levine's proposed alternative garners you no more usable data than heterophenomenology does. Moreover, if (b) outruns (a) — if you believe you have conscious experiences that you don't in fact have — then it is your beliefs that we need to explain, not the non-existent experiences! Sticking to the heterophenomenological standard, then, and treating (b) as the maximal set of primary data, is the way to avoid any commitment to spurious data.

---

[1] Doesn't interpretation require theory? Only in the minimal sense of presupposing that the entity interpreted is an intentional system, capable of meaningful communication. The task of unifying the interpretation of all the verbal judgments into a heterophenomenological world is akin to reading a novel, in contrast to reading what purports to be true history or biography. The issue of truth and evidence does not arise, and hence the interpretation is as neutral as possible between different theories of what is actually happening in the subject.

But what if some of your beliefs are inexpressible in verbal judgments? If you believe *that*, you can tell us, and we can add that belief to the list of beliefs in our primary data: 'S claims that he has ineffable beliefs about X'. If this belief is true, then we encounter the obligation to explain what these beliefs are and why they are ineffable. If this belief is false, we still have to explain why S believes (falsely) that there are these particular ineffable beliefs. As I put it in *Consciousness Explained*,

> You are *not* authoritative about what is happening in you, but only about what *seems* to be happening in you, and we are giving you total, dictatorial authority over the account of how it seems to you, about *what it is like to be you*. And if you complain that some parts of how it seems to you are ineffable, we heterophenomenologists will grant that too. What better grounds could we have for believing that you are unable to describe something than that (1) you don't describe it, and (2) confess that you cannot? Of course you might be lying, but we'll give you the benefit of the doubt (Dennett, 1991, pp. 96–7).

This is all quite obvious, but it has some under-appreciated implications. Exploiting linguistic communication in this way, you get a fine window into the subject's subjectivity but at the cost of a peculiar lapse in normal interpersonal relations. You *reserve judgment* about whether the subject's beliefs, as expressed in their communication, are true, or even well-grounded, but then you treat them as *constitutive* of that subject's subjectivity. (As far as I can see, this is the third-person parallel to Husserl's notion of bracketing or *epoché*, in which the normal presuppositions and inferences of one's own subjective experience are put on hold, as best one can manage, in order to get at the core experience, as theory-neutral and unencumbered as possible.) This interpersonal reserve can be somewhat creepy. To put it fancifully, suppose you burst into my heterophenomenology lab to warn me that the building is on fire. I don't leap to my feet and head for the door; I write down 'subject S believes the building is on fire'. 'No, really, it's on fire!' you insist, and I ask 'Would you like to expand on that? *What is it like* for you to think the building is on fire?' and so forth. In one way I am taking you as seriously as you could ever hope to be taken, but in another way I am not. I am not *assuming* that you are right in what you tell me, but just that that is what you do believe. Of course most of the data-gathering is not done by any such simple interview. Experiments are run in which subjects are prepared by various conversations, hooked up to all manner of apparatus, etc., and carefully debriefed. In short, heterophenomenology is nothing new; it is nothing other than the method that has been used by psychophysicists, cognitive psychologists, clinical neuropsychologists, and just about everybody who has ever purported to study human consciousness in a serious, scientific way.

This point has sometimes been misunderstood by scientists who suppose, quite reasonably, that since I am philosopher I must want to scold somebody for something, and hence must be proposing restrictions on standard scientific method, or discovering limitations therein. On the contrary, I am urging that the prevailing methodology of scientific investigation on human consciousness is not only sound, but readily extendable in non-revolutionary ways to incorporate *all* the purported exotica and hard cases of human subjectivity. I want to put the burden of proof on those who insist that third-person science is incapable of grasping the nettle of consciousness.

Let me try to secure the boundaries of the heterophenomenological method more clearly, then, since this has apparently been a cause of confusion. As Anthony Jack has said to me:

> It strikes me that heterophenomenology is a method in the same way that 'empiricism' is a method, but no more specific nor clearly defined than that. Given how general you seem to allow your definition of heterophenomenology to be, it is no surprise that everything conforms! Perhaps it would be clearer if you explained more clearly what it is supposed to be a counterpoint to — what it is that you object to. I know I am not the only one who has a feeling that you make the goalposts surprisingly wide. So what exactly is a foul? (Jack, personal correspondence).

Lone-wolf autophenomenology, in which the subject and experimenter are one and the same person, is a foul, not because you can't do it, but because it isn't science until you turn your self-administered pilot studies into heterophenomenological experiments. It has always been good practice for scientists to put themselves in their own experimental apparatus as informal subjects, to confirm their hunches about what it feels like, and to check for any overlooked or underestimated features of the circumstances that could interfere with their interpretations of their experiments. But scientists have always recognized the need to confirm the insights they have gained from introspection by conducting properly controlled experiments with naive subjects. As long as this obligation is met, whatever insights one may garner from 'first-person' investigations fall happily into place in 'third-person' heterophenomenology. Purported discoveries that cannot meet this obligation may inspire, guide, motivate, illuminate one's scientific theory, but *they* are not data — the beliefs of subjects about them are the data. Thus if some phenomenologist becomes convinced by her own (first-)personal experience, however encountered, transformed, reflected upon, of the existence of a feature of consciousness in need of explanation and accommodation within her theory, her *conviction that this is so* is itself a fine datum in need of explanation, by her or by others, but the truth of her conviction must not be presupposed by science.

Does anybody working on consciousness disagree with this? Does anybody think that one can take personal introspection by the investigator as constituting stand-alone evidence (publishable in a peer-reviewed journal, etc.) for any substantive scientific claim about consciousness? I don't think so. It is taken for granted, so far as I can see, by all the authors in this volume that there is no defensible 'first-person science' lying in this quarter, even though that would be the most obvious meaning of the phrase 'taking a first-person approach'. Thus Cytowic, and Hubbard and Jack, discuss the difficulties in confirming that synaesthesia is more or less what synaesthetes say it is, and never question the requirement that 'taking the phenomenological reports of these subjects seriously' (Hubbard and Jack, abstract) requires 'the personal interaction between subject and experimenter'. And when Hurlburt and Heavey say (abstract) 'For example, first-person investigators often rely on questions such as "What were you thinking when you . . . .?" or "How were you feeling when you. . . .?"' it apparently does not occur to them that these *aren't* first-person investigations; they are third-person investigations of the special kind that exploit the subject's capacity for verbal communication. They are heterophenomenological inquiries. So I think we can set aside lone-wolf autophenomenology in all its guises. It is not an attractive option, for familiar reasons. The experimenter/subject duality is not what is being challenged by those who want to go beyond the 'third-person' methodology. What other alternatives should we consider?

Several critics have supposed that heterophenomenology, as I have described it, is too agnostic or too neutral. Goldman (1997) says that heterophenomenology is not, as I claim, the standard method of consciousness research, since researchers 'rely substantially on subjects' introspective beliefs about their conscious experience (or lack thereof)' (p. 532). In personal correspondence (Feb 21, 2001, available as part of my debate with Chalmers, on my website, at http://ase.tufts.edu/cogstud/papers/chalmersdeb3dft.htm) he puts the point this way:

> The objection lodged in my paper [Goldman, 1997] to heterophenomenology is that what cognitive scientists *actually* do in this territory is not to practice agnosticism. Instead, they rely substantially on subjects' introspective beliefs (or reports). So my claim is that the heterophenomenological method is not an accurate description of what cognitive scientists (of consciousness) standardly do. Of course, you can say (and perhaps intended to say, but if so it wasn't entirely clear) that this is what scientists *should* do, not what they *do* do.

I certainly would play the role of reformer if it were necessary, but Goldman is simply mistaken; the adoption of agnosticism is so firmly built into practice these days that it goes without saying, which is perhaps why he missed it. Consider, for instance, the decades-long controversy about mental imagery, starring Roger Shepard, Steven Kosslyn, and Zenon Pylyshyn among many others. It was initiated by the brilliant experiments by Shepard and his students in which subjects were shown pairs of line drawings like the pair in figure 1, and asked to press one

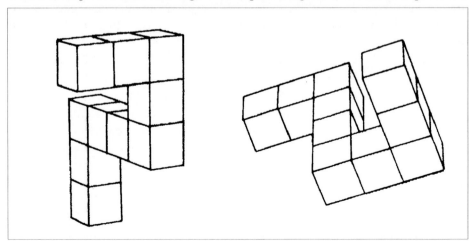

Figure 1

button if the figures were different views of the same object (rotated in space) and another button if they were of different objects. Most subjects claim to solve the problem by rotating one of the two figures in their 'mind's eye' or imagination, to see if it could be superimposed on the other. Were subjects really doing this 'mental rotation'? By varying the angular distance actually required to rotate the two figures into congruence, and timing the responses, Shepard was able to establish a remarkably regular linear relation between latency of response and angular displacement. Practiced subjects, he reported, are able to rotate such mental images at an angular velocity of roughly 60° per second (Shepard and Metzler, 1971). This didn't settle the issue, since Pylyshyn and others were quick to compose alternative hypotheses that could account for this striking temporal relationship. Further studies were called for and executed, and the controversy continues to generate new experiments and analysis today (see Pylyshyn, forthcoming, for an excellent survey of the history of this debate; also my commentary, Dennett, forthcoming, both in *Behavioral and Brain Sciences*). Subjects always *say* that they are rotating their mental images, so if agnosticism were not the tacit order of the day, Shepard and Kosslyn would have never needed to do their experiments to support subjects' claims that what they were doing (at least if described metaphorically) really was a process of image manipulation. Agnosticism is built into all good psychological research with human subjects. In psychophysics, for instance, the use of signal detection theory has been part of the canon since the 1960s, and it specifically commands researchers to control for the fact that the response criterion is under the subject's control although the subject is not himself or herself a reliable source on the topic. Or consider the voluminous research literature on illusions, both perceptual and cognitive, which standardly assumes that the data are what subjects judge to be the case, and never makes the mistake of 'relying substantially on subjects' introspective beliefs'.

The diagnosis of Goldman's error is particularly clear here: of course experimenters on illusions rely on subjects' introspective beliefs (as expressed in their judgments) about how it *seems* to them, but that *is* the agnosticism of heterophenomenology; to go beyond it would be, for instance, to assume that in size illusions there really were visual images of different sizes somewhere in subjects' brains (or minds), which of course no researcher would dream of doing.[2]

David Chalmers has recently made a similar, if vaguer, claim:

> Dennett . . . says scientists have to take a neutral attitude (taking reports themselves as data, but making no claims about their truth), because reports can go wrong. But this misses the natural intermediate option that Max Velmans has called critical phenomenology: accept verbal reports as a prima facie guide to a subject's conscious experience, except where there are specific reasons to doubt their reliability. This seems to be most scientists' attitude toward verbal reports and consciousness: it's not 'uncritical acceptance', but it's also far from the 'neutrality' of heterophenomenology (Chalmers, 2003).

---

[2] Goldman has responded to this paragraph in a series of emails to me, which I have included in an Appendix on the website mentioned above.

Chalmers neglects to say how Velmans' critical phenomenology is 'far from' the neutrality of heterophenomenology. I conducted a lengthy correspondence with Velmans on this score and was unable to discover what the purported difference is, beyond Velmans' insisting that his method 'accepts the reality of first-person experience', but since it is unclear what this means, this is something a good scientific method should be agnostic about. Neither Chalmers nor Velmans has responded to my challenge to describe an experiment that is licensed by, or motivated by, or approved by 'critical phenomenology' but off-limits to heterophenomenology, so if there is a difference here, it is one of style or emphasis, not substance. Chalmers has acknowledged this, in a way:

> Dennett 'challenges' me to name an experiment that 'transcends' the heterophenomenological method. But of course both views can accommodate experiments equally: every time I say we're using a verbal report or introspective judgment as a guide to first-person data, he can say we're using it as third-person data, and vice versa. So the difference between the views doesn't lie in the range of experiments 'compatible' with them. Rather, it lies in the way that experimental results are interpreted. And I think the interpretation I'm giving (on which reports are given prima facie credence as a guide to conscious experience) is by far the most common attitude among scientists in the field. Witness the debate about unconscious perception among cognitive psychologists about precisely which third-person measures (direct report, discrimination, etc.) are the best guide to the presence of conscious perception. Here, third-person data are being used as a (fallible) guide to first-person data about consciousness, which are of primary interest. On the heterophenomenological view, this debate is without much content: some states subserve report, some subserve discrimination, etc., and that's about all there is to say. I think something like this is Dennett's attitude to those debates, but it's not the attitude of most of the scientists working in the field (Chalmers, 2003).

Chalmers misconstrues my view, as we can see if we look more closely at a particular debate about unconscious perception, to see how heterophenomenology sorts out the issues. Consider *masked priming*. It has been demonstrated in hundreds of different experiments that if you present subjects with a 'priming' stimulus, such as a word or picture flashed briefly on a screen in front of the subject, followed very swiftly by a 'mask' — a blank or sometimes randomly patterned rectangle — before presenting the subjects with a 'target' stimulus to identify or otherwise respond to, there are conditions under which subjects will manifest behaviour that shows they have discriminated the priming stimulus, while they candidly and sincerely report that they were entirely unaware of any such stimulus. For instance, asked to complete the word stem *fri___*, subjects who have been shown the priming stimulus *cold* are more likely to comply with *frigid* and subjects who have been shown the priming stimulus *scared* are more likely to comply with *fright* or *frightened*, even though both groups of subjects claim not to have seen anything but first a blank rectangle followed by the target to be completed. Now are subjects to be trusted when they say that they were not conscious of the priming stimulus? There are apparently two ways theory can go here:

A. Subjects are conscious of the priming stimulus and then the mask makes them immediately forget this conscious experience, but it nevertheless influences their later performance on the target.
B. Subjects unconsciously extract information from the priming stimulus, which is prevented from 'reaching consciousness' by the mask.

Chalmers suggests that it is my 'attitude' that there is nothing to choose between these two hypotheses, but my point is different. It is open for scientific investigation to develop reasons for preferring one of these theoretical paths to the other, but *at the outset*, heterophenomenology is neutral, leaving the subject's heterophenomenological worlds bereft of any priming stimuli — that is how it seems to the subjects, after all — while postponing an answer to the question of how or why it seems thus to the subjects. Heterophenomenology is the beginning of a science of consciousness, not the end. It is the organization of the data, a catalogue of *what must be explained*, not itself an explanation or a theory. (This was the original meaning of 'phenomenology': a pretheoretical catalogue of the phenomena theory must account for.) And in maintaining this neutrality, it is actually doing justice to the *first-person* perspective, because you yourself, as a subject in a masked priming experiment, cannot discover anything in your experience that favours A or B. (If you think you can discover something — if you notice some glimmer of a hint in the experience, speak up! You're the subject, and you're supposed to tell it like it is. Don't mislead the experimenters by concealing something you discover in your experience. Maybe they've set the timing wrong for you. Let them know. But if they've done the experiment right, and you really find, so far as you can tell from your own first-person perspective, that you were not conscious of any priming stimulus, then say so, and note that both A and B are still options between which you are powerless to offer any further evidence.)

But now suppose scientists look for a good reason to favour A or B and find it. What could it be? A theory that could provide a good reason would be one that is well-confirmed in other domains or contexts and that distinguishes, say, the *sorts* of discriminations that can be made unconsciously from the sorts that require consciousness. If in this case the masked discrimination was of a feature that in all other circumstances could only be discriminated by a conscious subject, this would be a (fairly) good reason for supposing that, however it may be with other discriminations, in this case the discrimination was conscious-and-then-forgotten, not unconscious. Notice that if anything at all like this were discovered, and used as a ground for distinguishing A from B, it would be a triumph of *third-person* science, not due to anything that is accessible only to the subject's introspection. Subjects would *learn for the first time* that they were, or were not, conscious of these stimuli when they were taught the theory. It is the neutrality of heterophenomenology that permits such a question to be left open, pending further development of theory. And of course anyone proposing such a theory would have to have bootstrapped their way to their own proprietary understanding of what they meant by conscious and unconscious subjects, finding a consilience between our everyday assumptions about what we are conscious of and what we are not, on the one hand, and their own classificatory scheme on the other. Anything too extreme ('It turns out on our theory that most people are conscious for only a few seconds a day, and nobody is conscious of sounds at all; hearing is entirely unconscious perception') will be rightly dismissed as an abuse of common understanding of the terms, but a theory that is predictively fecund and elegant can motivate substantial abandonment of this anchoring lore. Only when such a theory is in place will we be able, for the first time, to *know what we mean* when we talk about 'the experiences themselves' as distinct from what we each, subjectively, take our experiences to be.

This sketches a clear path to settling the issue between A and B, or to discovering good reasons for declaring the question ill-posed. If Chalmers thinks that scientists do, and should, prefer a different attitude towards such questions, he should describe in some detail what it is and why it is preferable. In fact, I think that while there has been some confusion on this score (and some spinning of wheels about just what would count as favouring unconscious perception over conscious perception with forgetting), scientists are comfortable with the heterophenomenological standards.

Varela and Shear (1999) describe the empathy of the experimenter that they see as the distinguishing feature of a method they describe as first-person:

> In fact, that is how he sees his role: as an empathic resonator with experiences that are familiar to him and which find in himself a resonant chord. This empathic position is still partly heterophenomenological, since a modicum of critical distance and of critical evaluation is necessary, but the intention is entirely other: to meet on the same ground, as members of the same kind.... Such encounters would not be possible without the mediator being steeped in the domain of experiences under examination, as nothing can replace that first-hand knowledge. This, then, is a radically different style of validation from the others we have discussed so far (p. 10).

One can hardly quarrel with the recommendation that the experimenter be 'steeped in the domain of experiences' under examination, but is there more to this empathy than just good, knowledgeable interpretation? If so, what is it? In a supporting paper, Thompson speaks of 'sensual empathy', and opines: 'Clearly, for this kind of sensual empathy to be possible, one's own body and the Other's body must be of a similar type' (2001, p. 33). This may be clear to Thompson, but in fact it raises a highly contentious set of questions: Can women not conduct research on the consciousness of men? Can slender investigators not explore the phenomenology of the obese? Perhaps more to the point, can researchers with no musical training or experience ('tin ears') effectively conduct experiments on the phenomenology of musicians? When guidance from experts is available, one should certainly avail oneself of it, but the claim that one must *be* an expert (an expert musician, an expert woman, an expert obese person) before conducting the research is an extravagant one. Suppose, however, that it is true. If so, we should be able to discover this by attempting, and detectably failing, to conduct the research as well as the relevant experts conduct the research. *That* discovery would itself be something that could only be made by first adopting the neutral heterophenomenological method and then assaying the results in comparison studies. So once again, the neutral course to pursue is not to *assume* that men can't investigate the consciousness of women, etc., but to investigate the question of whether we can discover any good scientific reason to believe this. If we can, then we should adjust the standards of heterophenomenology accordingly. It is just common sense to design one's experiments in such a way as to minimize interference and maximize efficiency and acuity of data-gathering.

Is there, then, any 'radically different style of validation' on offer in these proposals? I cannot find any. Some are uneasy about the noncommital stance of the heterophenomenologist. Wouldn't the cultivation of deep trust between subject and experimenter be better? Apparently not. The history of *folie à deux* and Clever Hans phenomena suggests that quite unwittingly the experimenter and the subject may reinforce each other into artifactual mutual beliefs that evaporate when properly probed. But we can explore the question. It is certainly wise for the experimenter not to antagonize subjects, and to encourage an atmosphere of 'trust' — note the scare quotes. The question is whether experimenters should go beyond this and *actually trust* their subjects, or should instead (as in standard experimental practice) quietly erect the usual barriers and foils that keep subjects from too intimate an appreciation of what the experimenters have in mind. Trust is a two-way street, surely, and the experimenter who gets in a position where the subject can do the manipulating has lost control of the investigation.

I suspect that some of the dissatisfaction with heterophenomenology that has been expressed is due to my not having elaborated fully enough the potential resources of this methodology. There are surely many subtleties of heterophenomenological method that have yet to be fully canvassed. The policy of training subjects, in spite of its uneven history in the early days of psychology, may yet yield some valuable wrinkles. For instance, it might in some circumstances heighten the powers of subjects to articulate or otherwise manifest their subjectivity to investigators. The use of closed loop procedures, in which subjects to some degree control the timing and other properties of the stimuli they receive is another promising avenue. But these are not alternatives to heterophenomenology, which is, after all, just the conservative extension of standard scientific methods to data gathering from awake, communicating subjects.

Why *not* live by the heterophenomenological rules? It is important to appreciate that the reluctance to acquiesce in heterophenomenology as one's method is ideology-driven, not data-driven. Nobody has yet pointed to any variety of data that are inaccessible to heterophenomenology. Instead, they have objected 'in principle', perhaps playing a little gorgeous Bach for the audience and then asking the rhetorical question, 'Can anybody seriously believe that the wonders of human consciousness can be exhaustively plumbed by *third-person methods??*' Those who are tempted to pose this question should either temper their incredulity for the time being or put their money where their mouth is by providing the scientific world with some phenomena that defy such methods, or by describing some experiments that are clearly worth doing but that would be ruled out by heterophenomenology. I suspect that some of the antagonism to heterophenomenology is generated by the fact that the very neutrality of the methodology opens the door to a wide spectrum of theories, including some — such as my own — that are surprisingly austere, deflationary theories according to which consciousness is more like stage magic than black magic, requiring no revolution in either physics or metaphysics. Some opponents to heterophenomenology seem intent on building the mystery into the very setting of the problem, so that such deflationary theories are disqualified at the outset. Winning by philosophical footwork what ought to be won by empirical demonstration has, as Bertrand Russell famously remarked, all the advantages of theft over honest toil. A more constructive approach recognizes the neutrality of heterophenomenology and accepts the challenge of demonstrating, empirically, in its terms, that there are marvels of consciousness that cannot be captured by conservative theories.

References

Chalmers, David J. (2003), 'Responses to articles on my work' http://www.u.arizona.edu/~chalmers/responses.html#dennett2)
Churchland, Paul M. (1979), *Scientific Realism and the Plasticity of Mind* (Cambridge: Cambridge University Press).
Dennett, Daniel C. (1971), 'Intentional Systems', *J. Phil.*, **68**, pp. 87–106.
Dennett, Daniel C. (1982), 'How to study consciousness empirically, or Nothing comes to mind', *Synthese*, **59**, pp. 159–80.
Dennett, Daniel C. (1987), *The Intentional Stance* (Cambridge, MA: MIT Press/Bradford).
Dennett, Daniel C. (1991), *Consciousness Explained* (Boston, MA: Little Brown).
Dennett, Daniel C. (forthcoming), 'Does your brain use the images in it, and if so, how?' commentary on Pylyshyn (forthcoming).
Goldman, Alvin (1997), 'Science, Publicity and Consciousness', *Philosophy of Science*, **64**, pp. 525–45.
Levine, Joseph (1994), 'Out of the closet: A qualophile confronts qualophobia', *Philosophical Topics*, **22**, pp. 107–26.
Pylyshyn, Zenon W. (forthcoming), 'Mental Imagery: In search of a theory', Target article in *Behavioral and Brain Sciences*.
Shepard, R.N. and Metzler, J. (1971), 'Mental rotation of three-dimensional objects', *Science*, **171**, pp. 701–3.
Thompson, Evan (2001), 'Empathy and consciousness', *Journal of Consciousness Studies*, **8** (5–7), pp. 1–33.
Varela, Francisco and Shear, Jonathan (1999), 'First-person methodologies: What, Why, How?', *Journal of Consciousness Studies*, **6** (2–3), pp. 1–14.

Antoine Lutz and Evan Thompson

Neurophenomenology
Integrating Subjective Experience and Brain Dynamics in the Neuroscience of Consciousness

The paper presents a research programme for the neuroscience of consciousness called 'neurophenomenology' (Varela 1996) and illustrates it with a recent pilot study (Lutz et al., 2002). At a theoretical level, neurophenomenology pursues an embodied and large-scale dynamical approach to the neurophysiology of consciousness (Varela 1995; Thompson and Varela 2001; Varela and Thompson 2003). At a methodological level, the neurophenomenological strategy is to make rigorous and extensive use of first-person data about subjective experience as a heuristic to describe and quantify the large-scale neurodynamics of consciousness (Lutz 2002). The paper focuses on neurophenomenology in relation to three challenging methodological issues about incorporating first-person data into cognitive neuroscience: (i) first-person reports can be biased or inaccurate; (ii) the process of generating first-person reports about an experience can modify that experience; and (iii) there is an 'explanatory gap' in our understanding of how to relate first-person, phenomenological data to third-person, biobehavioural data.

I: Introduction

As this volume attests, a growing number of cognitive scientists now recognize the need to make systematic use of introspective phenomenological reports in studying the brain basis of consciousness (Jack and Shallice, 2001; Jack and Roepstorff, 2002; Dehaene and Naccache, 2001; Lutz *et al.*, 2002). Nevertheless, the integration of such first-person data into the experimental protocols of cognitive neuroscience still faces a number of epistemological and methodological challenges. The first challenge is that first-person reports can be biased or inaccurate (Nisbett and Wilson, 1977; Hurlbert and Heavey, 2001). The second challenge is that it seems reasonable to think that the very act or process of generating an introspective or phenomenological report about an experience can modify that experience. This challenge is closely related to broad conceptual and epistemological issues about the relationship of 'meta-awareness' to first-order experience (Schooler, 2002). Finally, there is the challenge of the so-called 'explanatory gap' in our understanding of how to relate (conceptually, methodologically and epistemologically) the first-person domain of subjective experience to the third-person domain of brain, body and behaviour (see Roy *et al.*, 1999). This gap still has to be adequately bridged, despite the presence of valuable neural models of consciousness and experimental evidence about the neural correlates of consciousness (or NCCs). As a result of these challenges, the status of first-person reports about experience remains a broad and problematic issue for cognitive science.

In this paper, we explore a research programme called 'neurophenomenology' that aims to make progress on these issues (Varela, 1996; 1997; 1999; Bitbol, 2002; Lutz, 2002; Rudrauf *et al.*, 2003). Neurophenomenology stresses the importance of gathering first-person data from phenomenologically trained subjects as a heuristic strategy for describing and quantifying the physiological processes relevant to consciousness. The general approach, at a methodological level, is (i) to obtain richer first-person data through disciplined phenomenological explorations of experience, and (ii) to use these original first-person data to uncover new third-person data about the physiological processes crucial for consciousness. Thus one central aim of neurophenomenology is to generate new data by incorporating refined and rigorous phenomenological explorations of experience into the experimental protocols of cognitive neuroscientific research on consciousness.

The term 'neurophenomenology' pays homage to phenomenological traditions in both Western philosophy (Spiegelberg, 1994; Petitot *et al.*, 1999) and Asian philosophy (Gupta, 1998; Wallace, 1998; Williams, 1998). ('Phenomenology' is capitalized in this paper when referring to the Western tradition derived from Edmund Husserl.) Phenomenology in this broad sense can be understood as the project of providing a disciplined characterization of the phenomenal invariants of lived experience in all of its multifarious forms. By 'lived experience' we mean experiences as they are lived and verbally articulated in the first-person, whether it be lived experiences of perception, action, memory, mental imagery, emotion, attention, empathy, self-consciousness, contemplative states, dreaming, and so forth. By 'phenomenal invariants' we mean categorical features of experience that are phenomenologically describable both across and within the various forms of lived experience. By 'disciplined characterization' we mean a phenomenological mapping of experience grounded on the use of 'first-person methods' for increasing one's sensitivity to one's own lived experience (Varela and Shear, 1999; Depraz *et al.*, 2003). The importance of disciplined, phenomenological examinations of experience for cognitive science was proposed and extensively discussed by Varela *et al.* (1991) as part of their 'enactive' approach to cognition. It was then subsequently elaborated by Varela (1996; 1997; 1999) into the specific research programme of neurophenomenology.

Of central importance to neurophenomenology is the employment of first-person phenomenological methods in order to obtain original and refined first-person data. It seems true both that people vary in their abilities as observers and reporters of their own experiences, and that these abilities can be enhanced through various phenomenological methods. 'First-person methods' are disciplined practices subjects can use to increase their sensitivity to their own experiences at various time-scales (Varela and Shear, 1999; Depraz *et al.*, 2003). These practices involve the systematic training of attention and self-regulation of emotion (see Section III). Such practices exist in phenomenology, psychotherapy and contemplative meditative traditions. Using these methods, subjects may be able to gain access to aspects of their experience (such as transient affective state or quality of attention) that otherwise would remain unnoticed and unavailable for verbal report. The experimentalist, on the other hand, using phenomenological reports produced by employing first-person methods, may be able to gain access to physiological processes that otherwise would remain opaque, such as the variability in brain response as recorded in EEG/MEG (see Lutz *et al.*, 2002 and Section V). Thus, at a methodological level, the neurophenomenological rationale for using first-person methods is to generate new data — both first-person and third-person — for the science of consciousness.

At an experimental level, the 'working hypothesis' of neurophenomenology (Varela, 1996) is that phenomenologically precise first-person data produced by employing first-person methods provide strong constraints on the analysis and interpretation of the physiological processes relevant to consciousness. Moreover, as Varela (1996; 1997; 1999) originally proposed, third-person data produced in this manner might eventually constrain first-person data, so that the relationship between the two would become one of dynamic 'mutual' or 'reciprocal constraints'. This means not only (i) that the subject is actively involved in generating and describing specific phenomenal invariants of experience, and (ii) that the neuroscientist is guided by these first-person data in the analysis and interpretation of physiological data, but also (iii) that the (phenomenologically enriched) neuroscientific analyses provoke revisions and refinements of the phenomenological accounts, as well as facilitate the subject's becoming aware of previously inaccessible or phenomenally unavailable aspects of his or her mental life (for a preliminary example, see Le Van Quyen and Petitmengin, 2002, on neurophenomenology as applied to the lived experience and neurodynamics of epileptic seizures).

To establish such reciprocal constraints, both an appropriate candidate for the physiological basis of consciousness and an adequate theoretical framework to characterize it are needed. Neurophenomenology is guided by the theoretical proposal (discussed in Section IV) that the best current candidate for the neurophysiological basis of consciousness is a flexible repertoire of dynamic large-scale neural assemblies that transiently link multiple brain regions and areas. This theoretical proposal is shared by a number of researchers, though specific models vary in their details (see Varela, 1995; Tononi and Edelman, 1998; Engel and Singer, 2001; Thompson and Varela, 2001). In this approach, the framework of dynamical systems theory is essential for characterizing the neural processes relevant to consciousness (see Le Van Quyen, 2003). In addition, neurophenomenology is guided by the 'embodied' approach to cognition (Varela *et al.*, 1991; Clark, 1997), which in its 'enactive' or 'radical embodiment' version holds that mental processes, including consciousness, are distributed phenomena of the whole active organism (not just the brain) embedded in its environment (Thompson and Varela, 2001, forthcoming; Varela and Thompson, 2003). These theoretical aspects of neurophenomenology have been presented extensively elsewhere (Varela, 1995; Thompson and Varela, 2001; Varela and Thompson, 2003; Rudrauf *et al.*, 2003), and are not the main focus of the present paper.

In summary, neurophenomenology is based on the synergistic use of three fields of knowledge:

1. (NPh1) First-person data from the careful examination of experience with specific first-person methods.
2. (NPh2) Formal models and analytical tools from dynamical systems theory, grounded on an embodied-enactive approach to cognition.
3. (NPh3) Neurophysiological data from measurements of large-scale, integrative processes in the brain.

In the following sections of this paper, we follow the steps of this threefold framework. In Section II we discuss some current concepts of consciousness as seen from a phenomenological perspective. In Section III we explain the basic features of first-person methods (NPh1). In Section IV we present the neurodynamical framework of neurophenomenology (NPh2 and NPh3). In Section V we review a pilot experimental study that illustrates the neurophenomenological approach. In Section VI we conclude by summarizing and discussing some of the implications of the neurophenomenological strategy for dealing with the challenge of integrating first-person data into cognitive neuroscience.

II: Concepts of Consciousness

A number of different concepts of consciousness can be distinguished in current research:

1. *Creature consciousness*: Consciousness of an organism as a whole insofar as it is awake and sentient (Rosenthal, 1997).
2. *Background consciousness* versus *state consciousness*: Overall states of consciousness, such as being awake, being asleep, dreaming, being under hypnosis, and so on (Hobson, 1999), versus specific conscious mental states individuated by their contents (Rosenthal, 1997; Chalmers, 2000). (The coarsest-grained state of background consciousness is sometimes taken to be creature consciousness (Chalmers, 2000).)

3. *Transitive consciousness* versus *intransitive consciousness*: Object-directed consciousness (consciousness-of), versus non-object-directed consciousness (Rosenthal, 1997).
4. *Access consciousness*: Mental states whose contents are accessible to thought and verbal report (Block, 2001). According to one important theory, mental contents are access-conscious when they are 'globally available' in the brain as contents of a 'global neuronal workspace' (Dehaene and Naccache, 2001; Baars, 2002).
5. *Phenomenal consciousness*: Mental states that have a subjective-experiential character (there is something 'it is like' for the subject to be in such a state) (Nagel, 1979; Block 2001).
6. *Introspective consciousness*: Meta-awareness of a conscious state (usually understood as a particular form of access consciousness) (Jack and Shallice, 2001; Hurlbert and Heavey, 2001; Jack and Roepstorff, 2002; Schooler, 2002).
7. *Pre-reflective self-consciousness*: Primitive self-consciousness; self-referential awareness of subjective experience that does not require active reflection or introspection (Wider, 1997; Williams, 1998; Gupta, 1998; Zahavi, 1999).

The relationships of these concepts to one another are unclear and currently the subject of much debate. A great deal of debate has centred on (4) and (5): Some theorists argue that it is possible for there to be phenomenally conscious contents that are inaccessible to thought, the rational control of action and verbal report (Block, 2001); others argue this notion of consciousness is incoherent, and hence deny the validity of the access/phenomenal distinction (Dennett, 2001).

This debate looks somewhat different when seen from a Phenomenological perspective. Central to this tradition, and to certain Asian phenomenologies (Gupta, 1998; Williams, 1998), are the notions of intentionality (which is related to (3) above) and pre-reflective self-consciousness (7). Pre-reflective self-consciousness is a primitive form of self-awareness believed to belong inherently to any conscious experience: Any experience, in addition to intending (referring to) its intentional object (transitive consciousness), is reflexively manifest to itself (intransitive consciousness).[1] Such self-manifesting awareness is a primitive form of self-consciousness in the sense that (i) it does not require any subsequent act of reflection or introspection but occurs simultaneously with awareness of the object; (ii) does not consist in forming a belief or making a judgment; and (iii) is 'passive' in the sense of being spontaneous and involuntary (see Zahavi and Parnas, 1998). A distinction is thus drawn between the 'noetic' process of experiencing, and the 'noematic' object or content of experience. Experience involves not simply awareness of its object (noema), but tacit awareness of

---

[1] It is important not to confuse this sense of 'reflexivity' (as intransitive and non-reflective self-awareness) with other usages that equate reflexivity and reflective consciousness (e.g. Block, 2001, who regards 'reflexivity' as 'phenomenality' plus 'reflection' or 'introspective access'). As Wider (1997) discusses in her clear and succinct historical account, the concept of reflexivity as pre-reflective self-awareness goes back to Descartes, and is a central thread running from his thought through Kant, Husserl, Sartre and Merleau-Ponty.

itself as process (noesis). For instance, when one consciously sees an object, one is also at the same time aware — intransitively, pre-reflectively and passively — of one's seeing; when one visualizes a mental image, one is thus aware also of one's visualizing. This tacit self-awareness has often been explicated as involving a form of non-objective bodily self-awareness — a reflexive awareness of one's 'lived body' (*Leib*) or embodied subjectivity correlative to experience of the object (Merleau-Ponty, 1962; Wider, 1997; Zahavi, 2002). Hence from a neurophenomenological perspective, any convincing theory of consciousness must account for this pre-reflective experience of embodied subjectivity, in addition to the object-related contents of consciousness (Varela *et al.*, 1991; Thompson and Varela, 2001; Zahavi, 2002).

Neurophenomenology thus corroborates the view, articulated by Panksepp (1998a,b) and Damasio (1999; Parvizi and Damasio, 2001), that neuroscience needs to explain both 'how the brain engenders the mental patterns we experience as the images of an object' (the noema in Phenomenological terms), and 'how, in parallel . . . the brain also creates a sense of self in the act of knowing . . . how each of us has a sense of "me" . . . how we sense that the images in our minds are shaped in our particular perspective and belong to our individual organism' (Parvizi & Damasio, 2001, pp. 136–7). In Phenomenological terms, this second issue concerns the noetic side of consciousness, in particular the noetic aspect of 'ipseity' or the minimal subjective sense of 'I-ness' in experience, which is constitutive of a 'minimal' or 'core self', as contrasted with a 'narrative' or 'autobiographical self' (Gallagher, 2000). As a number of cognitive scientists have emphasized, this primitive self-consciousness is fundamentally linked to bodily processes of life regulation, emotion and affect, such that all cognition and intentional action are emotive (Panksepp, 1998a, 1998b; Damasio, 1999; Watt, 1999; Freeman, 2000; Parvizi and Damasio, 2001), a theme central to Phenomenology (Merleau-Ponty, 1962; Jonas, 1966; Husserl, 2001).

This viewpoint bears on the access/phenomenal-consciousness debate as follows. According to Phenomenology, 'lived experience' comprises pre-verbal, pre-reflective and affectively valenced mental states (events, processes), which, while not immediately available or accessible to thought, introspection and verbal report, are intransitively 'lived through' subjectively, and thus have an experiential or phenomenal character. Such states, however, are (i) necessarily primitively self-aware, otherwise they do not qualify as conscious (in any sense); and (ii) because of their being thus self-aware, are access conscious in principle, in that they are the kind of states that can become available to thought, reflective awareness, introspection and verbal report, especially through first-person methods (see Section III).

In summary, whereas many theorists currently debate the access/phenomenal-consciousness distinction in largely static terms, neurophenomenology proposes to reorient the theoretical framework by emphasizing the dynamics of the whole noetic-noematic structure of consciousness, including the structural and temporal dynamics of the process of becoming reflectively or introspectively aware of experience, such that implicit and intransitively 'lived through' aspects of pre-reflective experience can become thematized and verbally described.

III: First-Person Methods

First-person methods are disciplined practices subjects can use to increase their sensitivity to their own experience from moment to moment (Varela and Shear, 1999). They involve systematic training of attention and emotional self-regulation. Such methods exist in Phenomenology (Depraz, 1999), psychotherapy (Gendlin, 1981; Epstein, 1996), and contemplative meditative traditions (Wallace, 1999). Some are routinely used in clinical and health programmmes (Kabat-Zinn et al., 1985), and physiological correlates and effects of these practices have been investigated (Austin, 1998; Davidson et al., 2003). The relevance of these practices to neurophenomenology derives from the capacity for attentive self-awareness they systematically cultivate. This capacity enables tacit, pre-verbal and pre-reflective aspects of subjective experience — which otherwise would remain simply 'lived through' — to become subjectively accessible and describable, and thus available for intersubjective and objective (biobehavioural) characterization.

First-person methods vary depending on the phenomenological, psychological or contemplative framework (Varela and Shear, 1999). We wish to underline certain common, generic operations of first-person methods. Of particular importance is the structural description of the disciplined process of becoming reflectively attentive to experience (Depraz et al., 2000, 2003). In Phenomenology, this disciplined process is known as the 'epoché' (Depraz, 1999). The epoché mobilizes and intensifies the tacit self-awareness of experience by inducing an explicit attitude of attentive self-awareness. The epoché has three intertwining phases that form a dynamic cycle (Depraz et al., 2000):

1. Suspension
2. Redirection
3. Receptivity

The first phase induces a transient suspension of beliefs or habitual thoughts about what is experienced. The aim is to 'bracket' explanatory belief-constructs in order to adopt an open and unprejudiced descriptive attitude. This attitude is an important prerequisite for gaining access to experience as it is lived pre-reflectively. The second phase of redirection proceeds on this basis: Given an attitude of suspension, the subject's attention can be redirected from its habitual immersion in the experienced object (the noema) towards the lived qualities of the experiencing process (the noetic act and its 'pre-personal' or 'pre-noetic' sources in the lived body).

During the epoché, an attitude of receptivity or 'letting go' is also encouraged, in order to broaden the field of experience to new horizons, towards which attention can be turned. Distinctions usually do not arise immediately, but require multiple variations. The repetition of the same task, for instance, enables new contrasts to arise, and validates emerging categories or invariants. Training is therefore a necessary component to cultivate all three phases, and to enable the emergence and stabilization of phenomenal invariants.

Downstream from this threefold cycle is the phase of verbalization or expression. The communication of phenomenal invariants provides the crucial step whereby this sort of first-person knowledge can be intersubjectively shared and calibrated, and related to objective data.

This explication of the procedural steps of the epoché represents an attempt to fill a lacuna of Phenomenology, which has often emphasized theoretical analysis and description, to the neglect of the pragmatics of the epoché as an embodied and situated act (Depraz, 1999). By contrast, the pragmatics of mindfulness-awareness (*shamatha-vipashyana*) in the Buddhist tradition are far more developed. This is one reason that the above description of the structural dynamics of becoming aware, as well as attempts to develop a more embodied and pragmatic phenomenology, have drawn from Buddhist traditions of mental cultivation (Varela *et al.*, 1991; Depraz *et al.*, 2000, 2003). One can also point to a recent convergence of theories and research involving introspection (Vermersch, 1999), the study of expertise and intuitive experience (Petitmengin-Peugeot, 1999), Phenomenology (Depraz, 1999) and meditative mental cultivation (Wallace, 1999). This convergence has also motivated and shaped the above description of the generic features of first-person methods (see Depraz *et al.*, 2000, 2003).

This stress on pragmatics represents an attempt to do justice to the difficulty of describing or reporting experiences as they are directly lived, rather than as they are assumed to be, either on the basis of a priori assumptions or extraneous theorizing. According to the Phenomenological way of thinking, in ordinary life we are caught up in the world and our various belief-constructs and theories about it. Phenomenologists call this unreflective stance the 'natural attitude'. The epoché aims to 'bracket' these assumptions and belief-constructs and thereby induce an open phenomenological attitude towards direct experience ('the things themselves'). The adoption of a properly phenomenological attitude is an important methodological prerequisite for exploring original constitutive structures and categories of experience, such as egocentric space, temporality and the subject-object duality, or spontaneous affective and associative features of the temporal flow of experience rooted in the lived body (for an overview of these topics, see Bernet *et al.*, 1993).[2]

The use of first-person methods in cognitive neuroscience clearly raises important methodological issues. One needs to guard against the risk of the experimentalist either biasing the phenomenological categorization or uncritically accepting it. Dennett (1991) introduced his method of

---

[2] Space prevents us from discussing the differences between Phenomenology and classical Introspectionism, particularly when the epoché is meant to initiate and sustain a form of transcendental philosophical reflection, by contrast with naturalistic investigation. Although we believe it is important to be clear about the differences between Phenomenology and Introspectionism (particularly because they are often lumped together by analytic authors), we also think that the historical debate between them is probably not directly relevant to the current renewal of interest in phenomenological and/or introspective evidence in cognitive science, as expressed in the contributions to this volume.

'heterophenomenology' (phenomenology from a neutral third-person perspective) in part as a way of guarding against these risks. His warnings are well taken. Neurophenomenology asserts that first-person methods are necessary to gather refined first-person data, but not that subjects are infallible about their own mental lives, nor that the experimentalist cannot maintain an attitude of critical neutrality. First-person methods do not confer infallibility upon subjects who use them, but they do enable subjects to thematize important but otherwise tacit aspects of their experience. Dennett to-date has not addressed the issue of the scope and limits of first-person methods in relation to heterophenomenology, so we are unsure where he stands on this issue. A full exchange on this issue would require discussion of the different background epistemological and metaphysical differences between Phenomenology and heterophenomenology concerning intentionality and consciousness. There is not space for such a discussion here.[3] We will therefore restrict ourselves to a comment about heterophenomenology as a method for obtaining first-person reports. Our view is that to the extent that heterophenomenology rejects first-person methods, it is too limited a method for the cognitive science of consciousness, because it is unable to generate refined first-person data. On the other hand, to the extent that heterophenomenology acknowledges the usefulness of first-person methods, then it is hard to see how it could avoid becoming in its practice a form of phenomenology, such that the prefix 'hetero' would become unnecessary.

Another methodological issue concerns the modification of experience by phenomenological training. It is to be expected that the stabilization of phenomenal categories through first-person methods will be associated with specific short-term or long-term changes in brain activity. It has been shown, for instance, that category formation during learning is accompanied by changes in the ongoing dynamics of the cortical stimulus representation (Ohl *et al.*, 2001). Yet the fact that phenomenological training can modify experience and brain dynamics is not a limitation, but an advantage. Anyone who has acquired a new cognitive skill (such as stereoscopic fusion, wine-tasting, or a second language) can attest that experience is not fixed, but dynamic and plastic. First-person methods help to stabilize phenomenal aspects of this plasticity so that they can be translated into descriptive first-person reports. As Frith writes in a recent comment on introspection and brain imaging: 'A major programme for 21st century science will be to discover how an experience can be translated into a report, thus enabling our experiences to be shared' (Frith, 2002). First-person methods help 'tune' experience, so that such translation and intersubjective corroboration can be made more precise and rigorous. The issue of the generality of data from trained subjects remains open, but seems less critical at this stage of our knowledge than the need to obtain new data about the phenomenological and physiological processes constitutive of the first-person perspective.

Frith, following Jack and Roepstorff (2002), also comments that 'sharing experiences requires the adoption of a second-person perspective in which a

---

[3] For discussion of this issue, see Thompson *et al.* (1999).

common frame of reference can be negotiated' (Frith, 2002). First-person methods help to establish such a reference frame by incorporating the mediating '-second-person' position of a trainer or coach. Neurophenomenology thus acknowledges the intersubjective perspective involved in the science of consciousness (Thompson, 2001). The subject needs to be motivated to cooperate with the experimentalist and empathetically to understand her motivations; and reciprocally the experimentalist needs to facilitate the subject's finding his own phenomenal invariants. Without this reciprocal, empathetically grounded exchange, there is no refined first-person data to be had.

IV: Neurodynamics (NPh2 and NPh3)

It is now widely accepted that the neural processes crucial for consciousness rely on the transient and ongoing orchestration of scattered mosaics of functionally specialized brains regions, rather than any single neural process or structure (Tononi and Edelman, 1998; Freeman, 1999; Dehaene and Naccache, 2001; Engel and Singer, 2001; Thompson and Varela, 2001). Hence a common theoretical proposal is that each cognitive or conscious moment involves the transient selection of a distributed neural population that is both highly integrated and differentiated, and connected by reciprocal, transient, dynamical links. A prelude to understanding the neural processes crucial for consciousness is thus to identify the mechanisms for large-scale brain processes, and to understand the causal laws and intrinsic properties that govern their global dynamical behaviours. This problem is known as the 'large-scale integration problem' (Varela *et al.*, 2001). Large-scale brain processes typically display endogenous, self-organizing behaviours (Engel *et al.*, 2001; Varela *et al.*, 2001), which are highly variable from trial to trial, and cannot be fully controlled by the experimentalist. Hence cognitive neuroscience faces at least a twofold challenge: (i) to find an adequate conceptual framework to understand brain complexity, and (ii) to relate brain complexity to conscious experience in an epistemologically and methodologically rigorous manner.

Brain complexity

For the first challenge, neurophenomenology endorses the strategy, now shared by many researchers, to use the framework of complex dynamical systems theory (Kelso, 1995; Freeman, 2001; Thompson and Varela, 2001; Varela *et al.*, 2001; Le Van Quyen, 2003). (This strategy corresponds to NPh2 above.) According to the dynamical framework, the key variable for understanding large-scale integration is not so much the individual activity of the nervous system's components, but rather the dynamic nature of the links among them. The neural counterpart of subjective experience is thus best studied not at the level of specialized circuits or classes of neurons (Crick and Koch, 1998), but through a collective neural variable that describes the emergence and change of patterns of large-scale integration (Varela *et al.*, 2001). Among the various ways to define this state variable, one recent approach is to use as a 'dynamical neural signature' the description and quantification of transient patterns of local and long-distance phase-synchronies occurring between oscillating neural populations at multiple frequency bands (Rodriguez *et al.*, 1999; Lutz *et al.*, 2002). (This proposal corresponds to a specific hypothesis that falls under the heading of NPh3 above.) The reasons for focusing on neural phase-synchrony are the evidence for its role as a mechanism of brain integration (Varela *et al.*, 2001), and its predictive power with respect to subsequent neural, perceptual and behavioural events (Engel *et al.*, 2001). Both animal and human studies demonstrate that specific changes in neural synchrony occur during arousal, sensorimotor integration, attentional selection, perception and working memory, which are all crucial for consciousness (for reviews and discussion, see Varela, 1995; Tononi and Edelman, 1998; Dehaene and Naccache, 2001; Engel and Singer, 2001; Engel *et al.*, 2001; Varela *et al.*, 2001). The irregularity and broad frequency band of these synchronies (2–80 Hz) suggest the presence of more complex forms of neural phase-synchronies than those investigated so far (Tass *et al.*, 1998; Rudrauf *et al.*, in press). Despite the need for theoretical development in these directions, the cornerstone assumption remains that large-scale, coherent neural activities constitute a fundamental self-organizing pole of integration in the brain, and that this pole provides a valuable physiological candidate for the emergence and the flow of cognitive-phenomenal states (Varela, 1995; Tononi and Edelman, 1998). To reveal the properties of this complex dynamic pole and the laws that govern it, several complementary temporal scales (10–100 milliseconds, 100–300 milliseconds, seconds, hours, days) and levels of description (neuronal, cell assembly, the whole brain) are probably needed. A cartography of conceptual and mathematical frameworks to analyse these spatio-temporal large-scale brain phenomena has been recently proposed (Le Van Quyen, 2003).

In addition, neurophenomenology favours an embodied approach to neural dynamics: The neurodynamic pole underlying the emergence and flow of cognitive-phenomenal states needs to be understood as necessarily embedded in the somatic contexts of the organism as a whole (the lived body in Phenomenological terms), as well as the environment (Thompson and Varela, 2001). In the case of human consciousness, the neurodynamic pole needs to be understood as necessarily embedded in at least three 'cycles of operation' constitutive of human life: (i) cycles of organismic regulation of the entire body; (ii) cycles of sensorimotor coupling between organism and environment; (iii) cycles of intersubjective interaction (for further discussion, see Thompson and Varela, 2001; Varela and Thompson, 2003).

In summary, neurophenomenology assumes that local and long-distance phase-synchrony patterns provide a plausible neural signature of subjective experience, and that the embodied-dynamical approach provides a theoretical language to specify cognitive acts in real time as cooperative phenomena at neural and organismic levels within and between brain, body and environment.

Relating brain complexity to conscious experience

Neurophenomenology's new and original methodological proposal, however, is to incorporate the experiential level into these neurodynamical levels in an explicit and rigorous way. The aim is to integrate the phenomenal structure of subjective experience into the real-time characterization of large-scale neural operations. The response to the second challenge is accordingly to create experimental situations in which the subject is actively involved in identifying and describing experiential categories that can be used to identify and describe dynamical neural signatures of experience. As will be seen when we discuss the pilot study, a rigorous relationship between brain complexity and subjective experience is thereby established, because original phenomenal categories are explicitly used to detect original neurodynamical patterns. Such joint collection and analysis of first-person and third-person data instantiates methodologically the neurophenomenological hypothesis that cognitive neuroscience and phenomenology can be related to each other through reciprocal constraints (Varela, 1996). The long-term aim is to produce phenomenological accounts of real-time subjective experience that are sufficiently precise and complete to be expressed in formal and predictive dynamical terms, which in turn could be expressed as specific neurodynamical properties of brain activity. Such twofold dynamical descriptions of consciousness could provide a robust and predictive way to link reciprocally the experiential and neuronal realms. We turn now to describe a pilot experimental study that illustrates the validity and fruitfulness of this research programme.

V: A Neurophenomenological Pilot Study

Background to the study

When an awake and alert subject is stimulated during an experiment, his brain is not idle or in a state of suspension, but is engaged in cognitive activity. The brain response derives from the interaction between this ongoing activity and the afferent stimulation that affects it (Engel *et al.*, 2001). Yet because this ongoing activity has not been carefully studied, most of the brain response is not understood. Successive exposure to the same stimulus elicits highly variable responses, and this variability is treated as unintelligible noise (and may be discarded by techniques that average across trials and/or subjects). The source of this variability is thought to reside mainly in fluctuations of the subjective cognitive context, as defined by the subject's attentional state, spontaneous thought processes, strategy to carry out the task, and so on. Although it is common to control, at least indirectly, for some of these subjective factors (such as attention, vigilance or motivation), the ongoing subjective mental activity has not yet been analysed systematically.

One strategy would be to describe in more detail this ongoing activity by obtaining verbal reports from human subjects. These reports should reveal subtle changes in the subject's experience, whether from trial to trial or across individuals.

This type of qualitative first-person data is usually omitted from brain-imaging studies, yet if methodological precautions are taken in gathering such data, they can be used to shed light on cognition via a joint analysis with quantitative measures of neural activity. Following this approach, a pilot neurophenomenological study (Lutz *et al.*, 2002) investigated variations in subjective experience for one limited aspect of visual perception, namely, the emergence of an illusory 3D figure during the perceptual fusion of 2D random-dot images with binocular disparities.

Experimental task

The task began with subjects fixating for seven seconds a dot pattern containing no depth cues. At the end of this 'preparation period', the pattern was changed to a slightly different one with binocular disparities. Subjects then had to press a button as soon as the 3D shape had completely emerged. Throughout the trial EEG signals were recorded, and immediately after the button-press subjects gave a brief verbal report of their experience. In these reports, they labelled their experience using phenomenal categories or invariants that they themselves had found and stabilized during the prior training session. The recording-session thus involved the simultaneous collection of first-person data (introspective/retrospective verbal reports) and third-person data (electrophysiological recordings and behavioural measures of button-pressing reaction time).

Training session

Subjects were intensively trained to perform the task in order to improve their perceptual discrimination and to enable them to explore carefully variations in their subjective experience during repeated exposure to the task. They were thus instructed to direct their attention to their own immediate mental processes during the task and to the felt-quality of the emergence of the 3D image.

This redirection of attention to the lived quality of experience corresponds to the epoché described in Section III. Its aim is to intensify the tacit self-awareness of experience by inducing a more explicit awareness of the experiencing process correlated to a given experiential content (the noetic-noematic structure of experience). More simply put, the aim is to induce awareness not simply of the 'what' or object-pole of experience (the 3D percept), but also of the necessarily correlated 'how' or act-pole of experience (the performance of perceptual fusion and its lived or subjective character). As described earlier, this method of becoming aware involves the three interlocking phases of suspension, redirection and receptivity.

In this pilot study, these phases were either self-induced by subjects familiar with them, or induced by the experimenter through open questions (Petitmengin-Peugeot, 1999). For example:

> Experimenter — 'What did you feel before and after the image appeared?'

Subject — 'I had a growing sense of expectation but not for a specific object; however, when the figure appeared, I had a feeling of confirmation, no surprise at all.'

or

'It was as if the image appeared in the periphery of my attention, but then my attention was suddenly swallowed up by the shape.'

Subjects were repeatedly exposed to the stimuli, and trial by trial they described their experience through verbal accounts, which were recorded on tape. In dialogue with the experimenters, they defined their own stable experiential categories or phenomenal invariants to describe the main elements of the subjective context in which they perceived the 3D shapes. The descriptive verbal reports from a total of four subjects were classified according to the common factor of the degree of preparation felt by the subject and the quality of his/her perception. This factor was used to cluster the trials into three main categories, described below: Steady Readiness, Fragmented Readiness and Unreadiness. Subcategories (describing the unfolding of the visual perception, for instance) were also found in individual subjects. They were not investigated in the pilot study.

1. Steady Readiness

In most trials, subjects reported that they were 'ready', 'present', 'here' or 'well-prepared' when the image appeared on the screen, and that they responded 'immediately' and 'decidedly'. Perception was usually experienced with a feeling of 'continuity', 'confirmation' or 'satisfaction'. These trials were grouped into a cluster SR, characterized by the subjects being in a state of 'Steady Readiness'.

2. Fragmented Readiness

In other trials, subjects reported that they had made a voluntary effort to be ready, but were prepared either less 'sharply' (due to a momentary 'tiredness') or less 'focally' (due to small 'distractions', 'inner speech' or 'discursive thoughts'). The emergence of the 3D image was experienced with a small feeling of surprise or 'discontinuity'. These trials formed a second cluster corresponding to a state of 'Fragmented Readiness'.

An intermediate cluster between these two clusters was defined for subject S3. This was described as a state of open attention without active preparation. It was unique to this subject who found that this state contrasted sharply with that of prepared Steady Readiness.

3. Unreadiness (Spontaneous Unreadiness, Self-Induced Unreadiness)

In the remaining trials, subjects reported that they were unprepared and that they saw the 3D image only because their eyes were correctly positioned. They were surprised by it and reported that they were 'interrupted' by the image in the middle of a thought (memories, projects, fantasies, etc.). This state of distraction occurred spontaneously for subjects S1 and S4, whereas S2 and S3 triggered it either by fantasizing or by thinking about plans (S3), or by visualizing a mental image (S2). To separate passive and active distraction, these trials were divided into two different clusters, Spontaneous Unreadiness for S1 and S4, and Self-Induced Unreadiness for S2 and S3.

Joint analysis of first-person data and third-person data

These phenomenal invariants found in the training session were used to divide the individual trials of the recording session into corresponding phenomenological clusters. The EEG signals were analysed to determine the transient patterns of local and long-distance phase-synchrony between oscillating neural populations, and separate dynamical analyses of the signals were conducted for each cluster. The phenomenological clusters were thus used as a heuristic to detect and interpret neural processes. The hypothesis was that distinct phenomenological clusters would be characterized by distinct dynamical neural signatures before stimulation (reflecting state of preparation), and that these signatures would then differentially condition the neural and behavioural responses to the stimulus. To test this hypothesis, the behavioural data and the EEG data were analysed separately for each cluster.

Results

By combining first-person data and the analysis of neural processes, the opacity in the brain responses (due to their intrinsic variability) is reduced and original dynamical categories of neural activity can be detected. For an example, we can consider the contrast between the two clusters of Steady Readiness and Spontaneous Unreadiness for one of the subjects (Figure 1 — see back cover). In the first cluster (A), the subject reported being prepared for the presentation of the stimulus, with a feeling of continuity when the stimulation occurred and an impression of fusion between himself and the percept. In the second cluster (B), the subject reported being unprepared, distracted, and having a strong feeling of discontinuity in the flux of his mental states when the stimulus was presented. He described a clear impression of differentiation between himself and the percept. These distinct features of subjective experience are correlated with distinct dynamical neural signatures (in which phase-synchrony and amplitude are rigorously separated in the dynamical analysis). During steady preparation, a frontal phase-synchronous ensemble emerged early between frontal electrodes and was maintained on average throughout the trial, correlating with the subject's impression of continuity. The average reaction time for this group of trials was short (300 ms on average). The energy in the gamma band (30–70 Hz) increased during the preparation period leading up to the time of stimulus presentation. This energy shift towards the gamma band occurred in all subjects and was specific to the 'prepared' clusters. The energy in the gamma band was always higher in anterior regions during the pre-stimulus period for subjects in the 'prepared' clusters than for subjects in the 'unprepared' clusters, whereas the energy in the slower bands was lower. These results suggest that the deployment of attention during the preparation strategy was characterized by an enhancement of the fast rhythms in combination with an attenuation of the slow rhythms. On the other hand, in the unprepared cluster, no stable phase-synchronous ensemble can be distinguished on average during the pre-stimulus period. When stimulation occurred, a complex pattern of weak synchronization and massive phase-scattering (desynchronization) between frontal and posterior electrodes was revealed. A subsequent frontal synchronous ensemble slowly appeared while the phase-scattering remained present for some time. In this cluster the reaction time was longer (600 ms on average). The complex pattern of synchronization and phase-scattering could correspond to a strong reorganization of the brain dynamics in an unprepared situation, delaying the constitution of a unified cognitive moment and an adapted response. This discontinuity in the brain dynamics was strongly correlated with a subjective impression of discontinuity.

Apart from these patterns common to all subjects, it was also found that the precise topography, frequency and time course of the synchrony patterns during the preparation period varied widely across subjects. These variations should not be treated as 'noise', however, because they reflect distinct dynamical neural signatures that remained stable in individual subjects throughout several recording sessions over a number of days (Figure 2 — see back cover).

Synopsis

This study demonstrated that (i) first-person data about the subjective context of perception can be related to stable phase-synchrony patterns measured in EEG recordings before the stimulus; (ii) the states of preparation and perception, as reported by the subjects, modulated both the behavioural responses and the dynamic neural responses after the stimulation; and (iii) although the precise shape of these synchrony patterns varied among subjects, they were stable in individual subjects throughout several recording sessions, and therefore seem to constitute a consistent signature of a subject's cognitive strategy or aptitude to perform the perceptual task. More generally, by using first-person methods to generate new first-person data about the structure of subjective experience, and using these data to render intelligible some of the opacity of the brain response, this pilot study illustrates the validity and fruitfulness of the neurophenomenological approach.

VI: Conclusion

This paper began by delineating three challenges faced by the attempt to integrate first-person data into the experimental protocols of cognitive neuroscience: (1) first-person reports can be biased or inaccurate; (2) introspective acts can modify their target experiences; and (3) there remains an 'explanatory gap' in our understanding of how to relate subjective experience to physiological and behavioural processes.

Neurophenomenology's strategy for dealing with the first two challenges is to employ first-person methods in order to increase the sensitivity of subjects to their own experience and thereby to generate more refined descriptive reports that can be used to identify and interpret third-person biobehavioural processes relevant to consciousness.[4] First-person methods intensify self-awareness so that it becomes less intrusive and more stable, spontaneous and fluid.

Such development implies that experience is being trained and reshaped. One might therefore object that one form of experience is replacing another, and hence the new experience cannot be used to provide insight into the earlier form of untrained experience. This inference, however, does not follow. There is not necessarily any inconsistency between altering or transforming experience (in the way envisaged) and gaining insight into experience through such transformation. If there were, then one would have to conclude that no process of cognitive or emotional development can provide insight into experience before the period of such development. Such a view is extreme and unreasonable. The problem with the objection is its assumption that experience is a static given, rather than dynamic, plastic and developmental. Indeed, it is hard to see how the objection could even be formulated without presupposing that experience is a fixed, predelineated domain, related only externally to the process of becoming aware, such that this process would have to supervene from outside, instead of being motivated by and called forth from within experience itself. First-person methods are not supposed to be a way of accessing such a (mythical) domain; they are supposed to be a way of enhancing and stabilizing the self-awareness already immanent in experience, thereby 'awakening' experience to itself.[5]

The final challenge comes from the 'explanatory gap'. We wish to draw a distinction between the 'explanatory gap' and the 'hard problem'. The 'explanatory gap' (in our usage) is the epistemological and methodological problem of how to relate first-person phenomenological accounts of experience to third-person cognitive-neuroscientific accounts. The 'hard problem' of consciousness is an abstract metaphysical problem about the place of consciousness in nature (Chalmers, 1996). It is standardly formulated as the issue of whether it is possible to derive subjective experience (or 'phenomenal consciousness') from objective physical nature. If it is possible, then physicalistic monism is supposed to gain support; if it is not possible, then property dualism (or substance dualism or idealism) is supposed to gain support.

---

[4] It is important to note that the biases and inaccuracies in first-person reports as documented in the well-known study by Nisbett and Wilson (1977) occurred when subjects made statements about what they took to be the causes of their mental states. Such statements, however, are not properly phenomenological, precisely because they engage in causal-explanatory theorizing beyond what is available to phenomenological description.

[5] Of course, the generality of results from phenomenologically trained subjects remains open as an empirical matter (as we indicated in Section III). Our point, however, is that this particular conceptual or a priori objection to first-person methods is misguided.

Although Varela (1996) proposed neurophenomenology as a 'methodological remedy for the hard problem', a careful reading of this paper indicates that neurophenomenology does not aim to address the metaphysical hard problem of consciousness on its own terms. The main reason, following analyses and arguments from phenomenological philosophers (Husserl, 1970; Merleau-Ponty, 1962), is that these terms — in particular the dichotomous Cartesian opposition of the 'mental' (subjectivist consciousness) versus the 'physical' (objectivist nature) — are considered to be part of the problem, not part of the solution. Space prevents further discussion of these issues here (see Bitbol, 2002, and Thompson and Varela, forthcoming, for discussion of neurophenomenology in relation to the hard problem).

With respect to the explanatory gap, on the other hand, neurophenomenology does not aim to close the gap in the sense of ontological reduction, but rather to bridge the gap at epistemological and methodological levels by working to establish strong reciprocal constraints between phenomenological accounts of experience and cognitive-scientific accounts of mental processes. At the present time, neurophenomenology does not claim to have constructed such bridges, but only to have proposed a clear scientific research programme for making progress on that task. Whereas neuroscience to-date has focused mainly on the third-person, neurobehavioural side of the explanatory gap, leaving the first-person side to psychology and philosophy, neurophenomenology employs specific first-person methods in order to generate original first-person data, which can then be used to guide the study of physiological processes, as illustrated in a preliminary way by the pilot study.

Our view is that the way experimental data are produced in the neuroscience of consciousness is implicitly shaped by the way the subject is mobilized in the experimental protocol. Experimental investigations of the neural correlates of consciousness usually focus on one or another particular noematic or noetic factor of experience, and accordingly (i) try to control as much as possible any variability in the content of subjective experience, and (ii) aim to minimize reliance on the subject's verbal reports. Yet this approach seems too limited, if the aim is to investigate the integrated, labile, self-referential and spontaneous character of conscious processes (see Varela, 1999; Hanna and Thompson, in press). Neurophenomenology, on the other hand, focuses on the temporal dynamics of the noetic-noematic structure as a whole. Thus, in the pilot study, the focus of investigation was the dynamics of the noetic-noematic interplay between the subjective-experiential context leading up to perception (hence the comparatively distant baseline of 7000 milliseconds: see Figure 1 — back cover), and the perceptual event itself. One aim of this sort of investigation is to understand the circular causality whereby (1) the antecedent and 'rolling' subjective-experiential context (noesis) can modulate the way the perceptual object appears (noema) or is experientially 'lived' during the moment of conscious perception, and (2) the content (noema) of this momentary conscious state can reciprocally affect the flow of experience (as noetic process). This global, noetic-noematic structure and its temporal dynamics are taken to reflect the endogenous, self-organizing activity of the embodied brain (Varela, 1999; Lutz, 2002), which in turn is understood as an autonomous dynamical system (Varela, 1995; Varela *et al.*, 2001; Rudrauf *et al.*, 2003). We believe that the most fruitful way for the experimentalist to investigate these sorts of processes, and to define and control the variables of interest, is to make rigorous and extensive use of the subject's first-person insight and descriptive verbal reports about her experience. Hence neurophenomenology, without denying the validity of trying to control experimentally the subjective context from the outside, favours a complementary 'endogenous' strategy that explicitly takes advantage of the first-person perspective in action. By thus enriching our understanding of both the first-person and third-person dimensions of consciousness, and creating experimental situations in which they reciprocally constrain each other, neurophenomenology aims to narrow the epistemological and methodological distance in cognitive neuroscience between subjective experience and brain processes.

To conclude this paper, let us point to a few general areas of research in which a neurophenomenological approach seems promising and complementary to more standard forms of biobehavioural and cognitive-scientific research (this list is meant to be illustrative, not exhaustive):

- Plasticity of human experience: To what extent can experience in domains such as attention, emotion, imagination and introspection be trained, and to what extent can such training modify structural and large-scale dynamical features of the human brain? (For a discussion of imagination in this context, see Varela and Depraz, 2003.)
- Time consciousness: Can phenomenological accounts of the different constitutive levels of time consciousness (Husserl, 1991) shed light on neurodynamics (Varela, 1999)? What role does emotion play in the spontaneous generation of the flow of consciousness (Freeman, 2000; Varela and Depraz, 2000)?
- Intersubjectivity: Can phenomenological accounts of intersubjectivity and empathy (Depraz, 1995) help to disentangle different aspects of intersubjective cognitive processes and their physiological basis (Thompson, 2001; Gallagher, 2003)?
- Dreaming: Can phenomenological explorations of dreaming through first-person methods of lucid dreaming (LaBerge, 1985, 1998, 2003) cast light on the neurodynamics of consciousness across sleeping, dreaming and wakefulness?

As we have proposed throughout this paper, the investigation of such empirical issues depends fundamentally on the ability of subjects to mobilize their insight about their experience and provide descriptive reports in a disciplined way compatible with the intersubjective standards of science. For this task, better procedural descriptions and pragmatics of the process of becoming aware of experience need to be developed (Varela and Shear, 1999; Depraz *et al.*, 2003). A paradigmatic neurophenomenological collaboration could therefore involve subjects with extensive training in the know-how of rigorous contemplative phenomenologies (such as those cultivated in Buddhist traditions), which seem to comprise stable experiential categories, detailed procedural descriptions and precise pragmatics, and have already begun to be explored in relation to cognitive science (Varela *et al.*, 1991; Austin, 1998; Wallace, 1998, 2003; Goleman *et al.*, 2003).

In Memoriam

We dedicate this paper to the memory of Francisco Varela, who first proposed the research programme of neurophenomenology (Varela, 1996), and profoundly shaped the ideas expressed here. For an obituary see: http://psyche.csse.monash.edu.au/v7/psyche-7-12-thompson.html

Acknowledgements

For helpful comments we thank Richard Davidson, Alva Noë, Andreas Roepstorff, Rebecca Todd, Dan Zahavi, and an anonymous reviewer for *JCS*. A.L. is supported by the Fyssen Foundation. E.T. is supported by the Social Sciences and Humanities Research Council of Canada through a Canada Research Chair, and by the McDonnell Project in Philosophy and the Neurosciences.

References

Austin, J. (1998) *Zen and the Brain* (Cambridge, MA: MIT Press).
Baars, B.J. (2002), 'The conscious access hypothesis: origins and recent evidence', *Trends in Cognitive Sciences*, 6, pp. 47–52.
Bernet, R. *et al.* (1993), *An Introduction to Husserlian Phenomenology* (Evanston, IL: Northwestern University Press).
Bitbol, M. (2002), 'Science as if situation mattered', *Phenomenology and the Cognitive Sciences*, 1, pp. 181–224.
Block, N. (2001), 'Paradox and cross purposes in recent work on consciousness', *Cognition*, 79, pp. 197–219.
Chalmers, D.J. (1996), *The Conscious Mind* (New York: Oxford University Press).
Chalmers, D.J. (2000), 'What is a neural correlate of consciousness?', in Metzinger (2000).
Clark, A. (1997), *Being There: Putting Brain, Body, and World Together Again* (Cambridge, MA: MIT/Bradford).
Crick, F. and Koch, C. (1998), 'Consciousness and neuroscience', *Cerebral Cortex*, 8, pp. 97–107.
Damasio, A.R. (1999), *The Feeling of What Happens: Body and Emotion in the Making of Consciousness* (London: Harcourt Brace).
Davidson, R. *et al.* (2003), 'Alterations in brain and immune function produced by mindfulness meditation', *Psychosomatic Medicine* (in press).
Dehaene, S. and Naccache, L. (2001), 'Towards a cognitive neuroscience of consciousness: basic evidence and a workspace framework', *Cognition*, 79, pp. 1–37.
Dennett, D.C. (1991), *Consciousness Explained* (Boston, MA: Little Brown).
Dennett, D.C. (2001), 'Are we explaining consciousness yet?', *Cognition*, 79, pp. 221–37.
Depraz, N. (1995), *Transcendence et incarnation: le statut de l'intersubjectivité comme altérité 'a soi chez Husserl* (Paris: Librarie Philosophique J. Vrin).
Depraz, N. (1999), 'The phenomenological reduction as *praxis*', in Varela and Shear (1999).
Depraz, N. *et al.* (2000), 'The gesture of awareness: an account of its structural dynamics', in *Investigating Phenomenal Consciousness*, ed. M. Velmans (John Benjamins Press).
Depraz, N. *et al.* (2003), *On Becoming Aware* (Amsterdam: John Benjamins Press).
Engel, A. and Singer, W. (2001), 'Temporal binding and the neural correlates of sensory awareness', *Trends in Cognitive Sciences*, 5, pp. 16–25.
Engel, A.K. *et al.* (2001), 'Dynamic predictions: oscillations and synchrony in top-down processing', *Nature Reviews Neuroscience*, 2, pp. 704–16.
Epstein, M. (1996), *Thoughts without a Thinker* (New York: Basic Books).

Freeman, W. (1999), 'Consciousness, intentionality, and causality', *Journal of Consciousness Studies*, **6** (11–12), pp. 143–72.
Freeman, W.J. (2000), 'Emotion is essential to all intentional behaviors', in *Emotion, Development, and Self-Organization*, ed. M. Lewis and I. Granic (Cambridge: Cambridge University Press), pp. 209–35.
Freeman, W. (2001), *Neurodynamics* (Berlin: Springer Verlag).
Frith, C. (2002), 'How can we share experiences?', *Trends in Cognitive Sciences*, **6**, p. 374.
Gallagher, S. (2000), 'Philosophical concepts of the self: implications for cognitive science', *Trends in Cognitive Sciences*, **4**, pp. 14–21.
Gallagher, S. (2003), 'Phenomenology and experimental design', *Journal of Consciousness Studies*, **10** (9–10), pp. 85–99.
Gendlin, E.T. (1981), *Focusing* (New York: Bantam).
Goleman, D. *et al.* (2003), *Destructive Emotions* (New York: Bantam).
Gupta, B. (1998), *The Disinterested Witness* (Evanston, IL: Northwestern University Press).
Hanna, R. and Thompson, E. (in press), 'Neurophenomenology and the spontaneity of consciousness', *Canadian Journal of Philosophy, Supplementary Volume*.
Hobson, J.A. (1999), *Consciousness* (New York: W.H. Freeman).
Hurlbert, R.T. and Heavey, C.L. (2001), 'Telling what we know: describing inner experience', *Trends in Cognitive Sciences*, **5**, pp. 400–3.
Husserl, E. (1970), *The Crisis of European Sciences and Transcendental Phenomenology*, trans. D. Carr (Evanston, IL: Northwestern University Press).
Husserl, E. (1991), *On the Phenomenology of the Consciousness of Internal Time (1893–1917)*, trans. J.B. Brough (Dordrecht: Kluwer Academic Publishers).
Husserl, E. (2001), *Analyses Concerning Passive and Active Synthesis. Lectures on Transcendental Logic*, trans. A.J. Steinbock (Dordrecht: Kluwer Academic Publishers).
Jack, A.I. and Roepstorff, A. (2002), 'Introspection and cognitive brain mapping: from stimulus-response to script-report', *Trends in Cognitive Sciences*, **6**, pp. 333–9.
Jack, A.I. and Shallice, T. (2001), 'Introspective physicalism as an approach to the science of consciousness', *Cognition*, **79**, pp. 161–96.
Jonas, H. (1966), *The Phenomenon of Life* (Chicago, IL: University of Chicago Press).
Kabat-Zinn J. *et al.* (1985), 'The clinical use of mindfulness meditation for the self-regulation of chronic pain', *Journal of Behavioral Medicine*, **8**, pp. 63–190.
Kelso, J.A.S. (1995), *Dynamic Patterns* (Cambridge, MA: MIT Press).
LaBerge, S. (1985), *Lucid Dreaming* (Los Angeles, CA: Tarcher).
LaBerge, S. (1998), 'Dreaming and consciousness', in *Toward a Science of Consciousness 2*, ed. S.R. Hameroff, A.W. Kaszniak and A.C. Scott (Cambridge, MA: MIT Press), pp. 495–504.
LaBerge, S. (2003), 'Lucid dreaming and the yoga of dream state', in Wallace (2003).
Le Van Quyen, M. and Petitmengin, C. (2002), 'Neuronal dynamics and conscious experience: an example of reciprocal causation before epileptic seizures', *Phenomenology and the Cognitive Sciences*, **1**, pp. 169–80.
Le Van Quyen, M. (2003), 'Disentangling the dynamic core: a research program for neurodynamics at the large scale', *Biological Research*, **36**, pp. 67–88.
Lutz, A. (2002), 'Toward a neurophenomenology as an account of generative passages: a first empirical case study', *Phenomenology and the Cognitive Sciences*, **1**, pp. 133–67.
Lutz, A. *et al.* (2002), 'Guiding the study of brain dynamics by using first-person data: synchrony patterns correlate with ongoing conscious states during a simple visual task', *Proceeedings of the National Academy of Sciences USA*, **99**, pp. 1586–91.
Merleau-Ponty, M. (1962), *Phenomenology of Perception*, trans. C. Smith (London: Routledge).
Metzinger, T. (ed. 2000), *Neural Correlates of Consciousness* (Cambridge, MA: MIT Press).
Nagel, T. (1979), 'What is it like to be a bat?', reprinted in *Mortal Questions*, ed. T. Nagel (Cambridge: Cambridge University Press).
Nisbett, R.E. and Wilson, T.D. (1977), 'Telling more than we can know: verbal reports on mental processes', *Psychological Review*, **84**, pp. 231–59.
Ohl, F.W. *et al.* (2001), 'Change in pattern of ongoing cortical activity with auditory category learning', *Nature*, **412**, pp. 733–6.
Parvizi, J. and Damasio, A. (2001), 'Consciousness and the brainstem', *Cognition*, **79**, pp. 135–59.
Panksepp, J. (1998a), *Affective Neuroscience: The Foundations of Human and Animal Emotions* (Oxford: Oxford University Press).
Panksepp, J. (1998b), 'The periconscious substrates of consciousness: affective states and the evolutionary origins of self', *Journal of Consciousness Studies*, **5** (5–6), pp. 566–82.
Petitot, J. *et al.* (ed. 1999), *Naturalizing Phenomenology* (Stanford, CA: Stanford University Press).
Petitmengin-Peugeot, C. (1999), 'The intuitive experience', in Varela and Shear (1999l.
Rodriguez, E. *et al.* (1999), 'Perception's shadow: long-distance synchronization of human brain activity', *Nature*, **397**, pp. 430–3.
Rosenthal, D.M. (1997), 'A theory of consciousness', in *The Nature of Consciousness*, ed. N. Block *et al.* (Cambridge, MA: MIT Press).

Roy, J.-M. *et al.* (1999), 'Beyond the gap: an introduction to naturalizing phenomenology', in Petitot *et al.* (1999).
Rudrauf, D. *et al.* (2003), 'From autopoiesis to neurophenomenology', *Biological Research*, **36**, pp. 27–66.
Schooler, J.W. (2002), 'Re-representing consciousness: dissociations between experience and meta-consciousness', *Trends in Cognitive Sciences*, **6**, pp. 339–44.
Spiegelberg, H. (1994), *The Phenomenological Movement* (Dordrecht: Kluwer Academic Publishers).
Tass, P. *et al.* (1998), 'Detection of n:m phase locking from noisy data: application to magneto-encephalography', *Physical Review Letters*, **81**, pp. 3291–4.
Thompson, E. (2001), 'Empathy and consciousness', *Journal of Consciousness Studies*, **8** (5–7), pp. 1–32.
Thompson, E. and Varela, F.J. (2001), 'Radical embodiment: neural dynamics and consciousness', *Trends in Cognitive Sciences*, **5**, pp. 418–25.
Thompson, E. and Varela, F.J. (forthcoming), *Radical Embodiment: The Lived Body in Biology, Cognitive Science, and Human Experience* (Cambridge, MA: Harvard University Press).
Thompson, E. *et al.* (1999), 'Perceptual completion: a case study in phenomenology and cognitive science', in Petitot *et al.* (1999).
Tononi, G. and Edelman, G.M. (1998), 'Consciousness and complexity', *Science*, **282**, pp. 1846–51.
Varela, F.J. (1995), 'Resonant cell assemblies: a new approach to cognitive functions and neuronal synchrony', *Biological Research*, **28**, pp. 81–95.
Varela, F.J. (1996), 'Neurophenomenology: A methodological remedy to the hard problem', *Journal of Consciousness Studies*, **3** (4), pp. 330–50.
Varela, F.J. (1997), 'The naturalization of phenomenology as the transcendence of nature', *Alter*, **5**, pp. 355–81.
Varela, F.J. (1999), 'The specious present: a neurophenomenology of time consciousness', in Petitot et. al. (1999).
Varela, F.J. and Depraz (2000), 'At the source of time: valence and the constitutional dynamics of affect', *Arob@se. Journal de lettre et de sciences humain*, **4** (1–2) http://www.arobase.to. Also published in *Ipseity and Alterity: Interdisciplinary Approaches to Intersubjectivity*, ed. S. Gallagher and S. Watson (Rouen: Presses Universitaires de Rouen, 2002).
Varela, F.J. and Depraz, N. (2003), 'Imagining: embodiment, phenomenology, and transformation', in Wallace (2003).
Varela, F.J. and Shear, J. (ed. 1999), *The View from Within* (Exeter: Imprint Academic).
Varela, F.J. and Thompson, E. (2003), 'Neural synchrony and the unity of mind: A neurophenomenological perspective', in *The Unity of Consciousness*, ed. A. Cleeremans (Oxford: Oxford University Press).
Varela, F.J. *et al.* (1991), *The Embodied Mind* (Cambridge, MA: MIT Press).
Varela, F.J. *et al.* (2001), 'The brainweb: phase synchronization and large-scale integration', *Nature Reviews Neuroscience*, **2**, pp. 229–39.
Vermersch, P. (1999), 'Introspection as practice', in Varela and Shear (1999).
Wallace, A. (1998), *The Bridge of Quiescence* (La Salle, IL: Open Court).
Wallace, B.A. (1999), 'The Buddhist tradition of *shamatha*: Methods for refining and examining consciousness', in Varela and Shear (1999).
Wallace, B.A. (ed. 2003), *Buddhism and Science: Breaking New Ground* (New York: Columbia University Press).
Watt, D.F. (1999), 'Emotion and consciousness: implications of affective neuroscience for extended reticular thalamic activating system theories of consciousness', Electronic publication of the Association for the Scientific Study of Consciousness. Available at: http://server.philvt.edu/assc/watt/default.htm
Wider, K.V. (1997), *The Bodily Nature of Consciousness: Sartre and Contemporary Philosophy of Mind* (Ithaca, NY: Cornell University Press).
Williams, P. (1998), *The Reflexive Nature of Awareness* (London: Curzon Press).
Zahavi, D. (1999), *Self-Awareness and Alterity* (Evanston, IL: Northwestern University Press).
Zahavi, D. (2002), 'First-person thoughts and embodied self-awareness: some reflections on the relation between recent analytic philosophy and phenomenology', *Phenomenology and the Cognitive Sciences*, **1**, pp. 7–26.
Zahavi, D. and Parnas, J. (1998), 'Phenomenal consciousness and self-awareness: a phenomenological critique of representational theory', *Journal of Consciousness Studies*, **5** (5–6), pp. 687–705.

Dan Zahavi and Josef Parnas

Conceptual Problems in Infantile Autism Research
Why Cognitive Science Needs Phenomenology

Until recently, cognitive research in infantile autism primarily focussed on the ability of autistic subjects to understand and predict the actions of others. Currently, researchers are also considering the capacity of autists to understand their own minds. In this article we discuss selected recent contributions to the theory of mind debate and the study of infantile autism, and provide an analysis of intersubjectivity and self-awareness that is informed both by empirical research and by work in the phenomenological tradition. This analysis uncovers certain problems in the theory-theory account of autism, and at the same time illustrates the potential value of phenomenology for cognitive science.

The aim of this article is to discuss, from a phenomenological perspective, selected recent contributions to the theory of mind debate and the study of infantile autism. A widespread view is that infantile autism does not only involve a lack of a theory of other minds, but also a lack of a theory of one's *own* mind. But how plausible is the assumption that intersubjectivity and self-consciousness involve a theory of mind? We will identify some problems in this explanatory model and in the final part of the article point to some alternative accounts found in phenomenology that might allow for a better understanding of the structure of intersubjectivity and self-awareness and thereby also for a better understanding of autism.

When speaking of phenomenology the vast majority of (Anglophone) philosophers and scientists are simply referring to a first-person description of what the 'what it is like' of experience is really like. In other words, there has been a tendency to identify phenomenology with some kind of introspectionism. But phenomenology is not just another name for a kind of psychological self-observation; rather it is the name of a philosophical approach specifically interested in consciousness and experience that was inaugurated by Husserl, and further developed and transformed by, among many others, Scheler, Heidegger, Fink, Gurwitsch, Sartre, Merleau-Ponty, Lévinas, and Henry.[1]

As we hope to show in the following, philosophical phenomenology can offer more than simply a compilation of introspective evidences, namely a conceptual framework for understanding subjectivity that might be of use for contemporary consciousness research.

I: The Theory of Mind Debate

'Theory of mind' is shorthand for our ability to attribute mental states to self and others and to interpret, predict and explain behaviour in terms of mental states. The theory of mind debate has mainly been a debate between two opposing views (although it has recently become more common to argue for some kind of mixed approach). On the one hand, we have the *theory-theory of mind*, and on the other the *simulation theory of mind*. The theory-theorists claim that the ability to explain and predict behaviour is underpinned by a folk-psychological theory dealing with the structure and functioning of the mind. We attribute beliefs to others by deploying theoretical knowledge. We come to understand others by using principles like the following: When an agent A acquires the belief that p and a rational thinker ought to infer q from the conjunction of p with other beliefs that A has, A comes to believe that q; or when an agent A attends to a situation S in a given way, and p is a fact about S perceptually salient in that way, then A acquires the belief that p (Botterill, 1996, pp. 115–6, cf. Carruthers, 1996a, p. 24). There is, however, disagreement among the theory-theorists about whether the theory in question is innate and modularized (Carruthers, Baron-Cohen), or whether it is acquired in the same way as scientific theories are acquired (Gopnik, Meltzoff). Most claim that there is some innate basis, but as Gopnik points out, it is necessary to distinguish between modularity nativism and starting-state nativism. The theory-formation theory, which takes the child to be a little scientist who is constructing and revising theories in the light of incoming data, can accept a certain nativism, but such initial structures are taken to be defeasible. They can and will be changed by new evidence (Gopnik, 1996, p. 171). Thus, for the theory-formation theory, there is a striking similarity between the acquisition of scientific knowledge and the child's increasing ability

---

[1] The phenomenologists themselves have vehemently denied that they should be engaged in some kind of introspective psychology. For some representative statements, cf. Gurwitsch, 1966, pp. 89–106; Husserl, 1984, pp. 201–16; Heidegger, 1993, pp. 11–17. One simple argument is that phenomenology must be appreciated as a kind of transcendental philosophy; it is not an empirical psychology. Another related argument is that introspection is typically understood as a mental operation that enables us to report and describe our own mental states. But strictly speaking phenomenology is not concerned with or based on operations of this kind. Rather, phenomenology is concerned with the phenomena, the appearances, their essential structures, and their conditions of possibility, and phenomenologists would typically argue that it is a metaphysical fallacy to locate these appearances within the mind, and to suggest that the way to access and describe them is by looking inside (*introspicio*) the mind. The entire facile divide between inside and outside is phenomenologically suspect, but this divide is precisely something that the term 'introspection' buys into and accepts (cf. Zahavi, 2003).

to adopt the intentional stance and mind-read, i.e., its ability to interpret behaviour in terms of an agent's mental state. It is basically the same cognitive processes that are responsible for scientific progress and for the development of the child's understanding of the mind (Gopnik, 1996, p. 169). In contrast, the modularists claim that the core of the folk-psychological theory is hardwired. As they point out, if the theory were merely the product of a scientific investigation, why is it then culturally universal and why do all children reach the same theory at the same age (Carruthers, 1996a, p. 23)? The theory is forged by evolution and innately given, and although it might need experience as a trigger, the theory of mind module will not be modified by experience.

Whereas the theory-theorists claim that we employ a theory about the psychological when we predict and explain the behaviour of others, the simulationists claim that we possess no such theory, or at least none complete enough, to underpin all our competence with psychological notions (Heal, 1996, p. 75). Whereas the theory-theorists make use of what Gordon has called a cold methodology, and argues that our understanding of others chiefly engages intellectual processes, moving by inference from one belief to the other, and making no use of our emotions, motivations and practical reasoning, the simulationists employ a hot methodology and argue that our understanding of others exploits our own motivational and emotional resources and our own capacity for practical reasoning (Gordon, 1996, p. 11). Thus, in contrast to the theory-theorists, the simulationists would argue that what lies at the root of our mindreading abilities is an ability to project ourselves imaginatively into another person's perspective, simulating their mental activity with our own.

In the following, our focus will be on the theory-theory of mind.[2] What exactly do the theory-theorists mean by theory? As we have already seen, the views differ. Some take the theory of mind to be a theory in a very literal sense and compare it to a scientific theory.[3] Others take it in a more extended sense and compare it to a set of rules of symbol manipulation instantiated in an innate module. Some take the theory in question to be explicit, to be something the agent is conscious of, others take it to be more or less implicit and tacit, and to be something that operates on a subpersonal level.

Generally speaking, however, many theory-theorists have tended to construe theory in a rather loose sense in order to increase the plausibility of their own position, but the danger they have thereby run is to make the notion of theory employed completely vacuous. In the end, everything would turn out to be theoretical, including cooking, gardening and fishing. In order to avoid this, some theory-theorists have simply bitten the bullet, and have accepted a strong definition of theory that entails much more than simply some kind of semantic holism.

[2] This is not to say, however, that the phenomenological account can simply be identified with the existing simulationist approach. From a phenomenological point of view there are problems with the simulation theory as well, not the least its reliance on some kind of argument from analogy seems problematic (for an extensive criticism of the argument from analogy, cf. Avramides, 2001).

[3] Gopnik and Wellman have compared the transition that occurs between the three-year-old and the four-year-old child's understanding of mind to the transition between Copernicus' *DRevolutionibus* and Kepler's discovery of elliptical orbits (Gopnik & Wellman, 1995, p. 242).

Botterill, a theory-theorist, takes both explanatory and predictive power as well as counterfactual projection to be necessary features of any theory. He also mentions the introduction of unobservable entities and the implicit definition of concepts as a recurrent (but non-necessary) element, and finally argues that theories are characterized by producing cognitive economy through the integration of information in a small number of general principles (Botterill, 1996, pp. 107–9).

The theory-theory claims that mastery of mental concepts is constituted by knowledge of a psychological theory. More specifically, our understanding of mental notions depends upon our knowledge of the positions that these notions occupy within the theory. Thus, the notions are taken to receive their sense from the theory in which they are embedded, rather than through some demonstrative identification or direct acquaintance. This is probably one of the most characteristic features of the theory-theory: It denies that our reference to mental states such as beliefs and desires is derived from our direct experience of such mental states, and instead argues that these concepts are theoretical postulates that have been developed through a process of abstract theorizing. Thus, since the theory-theorists frequently consider the reference to mental states to be a question of an inference to best explanation and prediction of behavioural data, they generally take mental states to be unobservable and theoretically postulated entities.[4]

II: Infantile Autism and the Theory-theory Account of Self-awareness

According to many theory-theorists (Gopnik, Carruthers, Frith and Happé) — at least if one takes some of their explicit statements at face value — we come to know our own beliefs and occurrent mental states just like we come to know the beliefs and experiences of others. In both cases, the same cognitive mechanism is in use, in both cases we are dealing with a process of mindreading, in both cases we are dealing with the application of a theory of mind. Thus, according to what might be labelled the theory-theory account of self-awareness, my access to my own mind depends on the same mechanisms that I use in attributing mental states to others. In both cases the access, the understanding, and the knowledge is theory mediated, and the mediating theory is the same for self and for other (Carruthers & Smith, 1996, p. 3; Gopnik, 1993, p. 3; Frith & Happé, 1999, p. 7). Even though we seem to perceive our own mental states directly, this direct perception is an illusion.

If this view is correct, there should be no fundamental difference in the development of our ability to attribute mental states to self and other, since the same cognitive mechanism would supposedly be used in both cases.[5] But what are the empirical findings? Do they support the existence of such a parallelism? Let us take a look at one of the most well-known type of tests: the *false-belief* task.

---

[4] That theory-theorists really do think that belief, desires and other mental states are unobservable entities is vividly illustrated by the following passage taken from Tooby's and Cosmides' foreword to Baron-Cohen's *Mindblindness*: 'Normal humans everywhere not only "paint" their world with colour, they also "paint" beliefs, intentions, feelings, hopes, desires and pretenses onto agents in their social world. They do this despite the fact that no human has ever seen a thought, a belief or an intention'(Baron-Cohen, 1995, p. xvii).

In the so-called *Smarties task* children are shown a candy box. Based on its appearance, children first believe that the box contains sweets, but the box is then opened and is shown to contain pencils. The box is then closed again, and the children are asked what other children, who have not yet seen inside the box, will think it contains. The average four-year-old will answer that other children will think it contains candy, whereas younger children will answer pencils. They apparently fail to understand that other persons' beliefs may be false (Frith & Happé, 1999, pp. 3–4).

Why this interest in children's ability to succeed on the false-belief task? Supposedly because the child, in order to succeed, must be able to contrast its own belief about the world with the belief of another, i.e., it must be able to have beliefs about beliefs, it must be in possession of a theory of mind. Moreover, in order to ascribe false beliefs to others (and to itself) it must understand that our beliefs might differ from reality. It must be in possession of an awareness of the difference between reality and our beliefs about reality. In order to test the existence of a parallelism in the attribution of mental states to self and to other, a variation of the *Smarties task* was devised. The children were presented with the deceptive candy box that was full of pencils. They were first asked the above-mentioned questions, and then the following question was added: 'When you first saw the box, before we opened it, what did you think was inside it?' Somewhat surprisingly, one-half to two-thirds of the three-year-olds said that they had originally thought that it contained pencils. They apparently failed to remember their own past false beliefs, i.e., three-year-old children apparently have as much trouble understanding their own past false beliefs as they have in understanding the false beliefs of others (Gopnik, 1993, pp. 6–8). According to Gopnik, this finding seriously challenges our commonsense intuition that our access to our own mental states is of a quite different (and far more direct and immediate) nature than our access to the mental states of others. In fact, when children can report and understand the psychological states of others, they can report having had those states themselves, and when they cannot report and understand the psychological states of others, they report that they have not had those states themselves (Gopnik, 1993, p. 9). In short, there is little evidence that mental states are attributed to self before they are attributed to others, and vice versa.

If our acquisition of beliefs about our own mental states parallels our acquisition of beliefs about the mental states of others, and if the epistemic source is fundamentally the same in both cases, why do we normally tend to believe that there is such a big difference between the two? The explanation offered by both Gopnik and Carruthers is that we have become experts on reading our own mind, and after having reached a certain expertise we tend to see things at once, although what we see is actually the result of a complex theoretical process. We draw on an accumulated theoretical knowledge, but our expertise makes us unaware of the inferential processes and makes us believe that our experience is

---

[5] Occasionally, some of the theory-theorists have been cautious enough to admit that this parallelism might not hold true for all kinds of mental states, but there is no general agreement about what should count as the relevant exceptions.

immediate and non-inferential. In other words, self-knowledge or self-consciousness can be thought of in analogy with the theory-laden perception of theoretical entities in science. Just as a diagnostician can sometimes see a cancer in the blur of an X-ray picture, so, too, each of us can sometimes see that we are in a state accorded such-and-such a role by folk-psychological theory (Gopnik, 1993, p. 11; Carruthers, 1996a, p. 26; 1996b, pp. 259–60).

It is now time to turn to infantile autism. The study of autism has been central to the theory of mind debate ever since it was suggested that autists have problems on theory of mind tasks. A dominant explanation, favoured by, for instance, Uta Frith, Alan Leslie and Simon Baron-Cohen, is that three of the cardinal symptoms in infantile autism, the so-called classical triad — impairment in social interaction and empathy, in verbal and non-verbal communication, and lack of creativity and imagination (including paucity in pretend play) — are the result of a failure in the development of the capacity to mind-read (Baron-Cohen *et al.*, 1985). More specifically, some of the core features in autism have been interpreted as the result of a mind-blindness, namely as an inability to understand mental states, and this selective impairment has been explained by reference to a damaged or destroyed theory of mind mechanism.

As we have just seen, theory-theorists typically argue that the theory of mind is involved not only in the understanding of other people's mental states, but also in the understanding of one's own mental states. If this is true, it has some rather obvious consequences for the understanding of infantile autism. If autists lack a theory of mind, and if a theory of mind is required for self-awareness, then autists should be 'as blind to their own mental states as they are to the mental states of others'(Carruthers, 1996b, p. 262; cf. Frith & Happé, 1999, pp. 1, 7). If by hypothesis it is a theory of mind that provides the network of theoretical concepts and principles which enables us to understand mental states, the autist will either be unable to recognize his own mental states as such at all, or he will only be able to do so through a slow and painstaking learning process (Carruthers, 1996b, p. 260; Frith & Happé, 1999, p. 2). Thus, Frith and Happé are basically proposing that persons with autism can only judge their own mental states by their actions (Frith & Happé, 1999, p. 11), i.e., that autists do not have direct, immediate, or non-inferential access to their own mind, but only know what transpires in their mind by applying a theory of mind to their own behaviour.

Are there any experimental data in support of this hypothesis? Both Baron-Cohen and Frith and Happé have argued that one can test the presence of self-awareness in infantile autists by using classical theory of mind tasks, such as the false-belief task or the appearance-reality task. Thus, even far older autistic children with good general cognitive capacities and high IQ scores answer the way normal three-year-olds do when confronted with theory of mind tasks. (By contrast, non-autistic children with other mental handicaps perform relatively well on these tasks). When confronted with the *Smarties task* autists have as much trouble remembering (self-ascribing) their own recent false beliefs as they have in attributing false beliefs to other people. Similarly, autistic children fail on the so-called *appearance-reality* task. In this task, children are presented with an object that appears to be one thing, but which in reality is something quite different. Typically, a painted sponge that looks just like a rock has been used. The children are then asked what that object looks like, and what it really is. Whereas normal four-year-olds are able to answer correctly, three-year-olds typically give the same answer to both questions: The object looks like a sponge, and it is a sponge (or the object looks like a rock and really is a rock) (Gopnik, 1993, p. 4; Baron-Cohen, 1989, p. 591). In an experiment, infantile autists (mean verbal age = 8.48) were confronted with plastic chocolates. They thought that the objects both looked like chocolate and really were chocolate. In fact, in one setting, the autistic children persisted in trying to eat the fakes long after having discovered their plastic quality (Baron-Cohen, 1989, p. 594). Since the ability to understand the appearance-reality distinction is taken to entail the ability to attribute mental states to oneself, this failure has in turn been interpreted as suggesting that the autists lack awareness of their own mental states (Baron-Cohen, 1989, p. 596; Carruthers, 1996b, p. 260).

III: Conceptual Issues

That the study of infantile autism is not merely an empirical enterprise but rather something that necessitates a good deal of conceptual analysis should be evident by now. In the following, we wish to point to some conceptual problems that in our view characterize the arguments by Gopnik, Baron-Cohen, and Frith and Happé. To anticipate our general criticism: There seems to be a gap between the experimental data and the theoretical claims and conclusions made by these authors. Moreover, their argumentations are typically based on a conception of both self-consciousness and intersubjectivity that are open to debate.

1. Gopnik

When reading her article 'How we know our minds: The illusion of first-person knowledge of intentionality', one occasionally gets the impression that Gopnik wants to challenge first-person authority as such. Thus, in the beginning of the article, she makes some rather far-reaching statements, saying, for instance, that developmental evidence questions the commonly held belief that the process of discovering our own psychological states is fundamentally different from the process of discovering someone else's states (Gopnik, 1993, pp. 1, 9; cf. Gopnik *et al.*, 2001, p. 47). On closer inspection, it becomes clear, however, that Gopnik is actually prepared to distinguish between different types of mental states, and that her main target is the notion of intentionality. In contrast to what might be the case for simple sensations (Gopnik 1993, pp. 2, 10), Gopnik takes intentionality to be a defeasible theoretical construct, and she argues that my beliefs about the intentionality of my own mental states share the same cognitive history as my beliefs about the intentionality of mental states belonging to others. As she writes,

> We first have psychological states, then we observe the behaviours and the experiences that they lead to in ourselves and others, then we construct a theory about the causes of

those behaviours and experiences that postulate intentionality, and *only* then do we have an experience of the intentionality of those states (Gopnik, 1993, p. 12).

The data provided by Gopnik is supposed to confirm the existence of a crucial parallelism in the attribution of intentional states to others and in the attribution of intentional states to self. However, the most significant of the findings presented by Gopnik demonstrates the existence of a parallelism in the attribution of *current* false beliefs to others and in the attribution of *past* false beliefs to self, but it is rather unclear why these findings — puzzling and interesting as they are — should warrant the kind of general claim made by Gopnik.

Apparently, however, Gopnik's idea is that unless you are able to appreciate that it is possible to have mistaken beliefs, you cannot understand what it means to have beliefs or intentional states at all (Gopnik, 1993, p. 6). It is certainly reasonable to assume that if a child can understand what a false belief is, then it can also understand what a belief is. But is it also reasonable to conclude that unless a child can understand false beliefs it cannot understand beliefs? Certainly, if we are talking of a full-fledged theoretical understanding of beliefs, i.e., of an actual theory of beliefs, such a theory must involve an understanding of and explanation of the possibility of error. If it did not, we would typically say that it was not really a theory of beliefs, or at best, that it was a very inadequate theory of beliefs. However, it is hardly surprising that we have high requirements for what a theory should entail. The question is whether or not it is appropriate to apply the same strong requirements to an infant, and to claim that the infant does not understand beliefs, in the sense of experiencing intentional (object-directed) states, i.e., in the sense of consciously entertaining beliefs, unless it is capable of attributing false beliefs to self and others.

In a recent article, Nichols and Stich have launched a strong attack on the theory-theory of self-awareness. As they point out, there are basically three ways to interpret the theory. First of all, it could be taken to involve the claim that the only information we have about our own mental states is the kind of evidence that others are also in possession of. In this sense, knowledge of self and knowledge of others would be completely analogous. However, as they point out, this is a form of pure behaviourism that is very implausible. Secondly, the theory could be taken to involve the concession that my access to my own mental states is based on information that is not available in the case of my access to the mental states of others. The problem, however, is that the theory never spells out what exactly this information is. Gopnik refers to first-person psychological experience as 'the Cartesian buzz' (Gopnik, 1993, p. 11), but as Nichols and Stich point out, this is not exactly an illuminating answer. Finally, the theory-theory of self-awareness might argue that the additional information that is available in my own case is information about my own mental states. But as they then point out, if this information is available to me from the outset, there is no reason to introduce and involve any theory of mind mechanism (Nichols & Stich, 2002, p. 12).

2 Baron-Cohen

In his article 'Are autistic children "behaviorists"?' Baron-Cohen suggests that autists who lack a theory of mind and who are incapable of adopting the intentional stance will be forced to view the world in terms of behavioural and physical events (Baron-Cohen, 1989, p. 580). Baron-Cohen presents different experimental data and he writes that autistic children are unable to distinguish mental entities from physical entities, and that they simply seem unaware of the special nature of mental entities. One of the examples concerns dreaming. Autistic children were asked where dreams are located, in the room or inside their head. Most would say in the room, and as Baron-Cohen writes, this finding suggests that autistic children simply do not understand the *concept* 'dream' (Baron-Cohen, 1989, p. 587). Baron-Cohen also argues that autists do not understand linguistic terms referring to thinking, pretending, etc., and that their inability to distinguish appearance from reality reflects a lack of metacognition or metarepresentation (Baron-Cohen, 1989, p. 596). All of these statements suggest that autists suffer from some defect in higher cognition. It is their conceptual and linguistic skills that are impaired and defective. At the same time, however, Baron-Cohen repeatedly makes some assertions that can be construed as involving a much stronger claim, saying in turn that autistic children 'appear oblivious to the existence of mental phenomena'(Baron-Cohen, 1989, p. 588); that 'autistic subjects [. . .] appear to be largely unaware of the mental world' (Baron-Cohen, 1989, p. 590); and that autistic subjects are 'unaware of their own mental states'(Baron-Cohen, 1989, p. 595).

Baron-Cohen concludes the article by denying that all internal (mental) states are beyond the comprehension of autists. In his view, they do, in fact, have an intact understanding of desire and of simple emotions like happiness, sadness and hunger. Thus, rather than saying that autists are behaviourists, it would be better to call them 'desire-psychologists'. That is, autists may be able to predict people's actions in terms of desires and physical causes, but they fail to develop a belief-desire psychology (Baron-Cohen, 1989, pp. 597–8). It is rather unclear to us how Baron-Cohen can reconcile the claim that autists understand certain emotions with his claim that they are oblivious to mental phenomena.

As we have seen, Baron-Cohen takes the appearance-reality task as a way of testing infants' ability to represent the distinction between their perception of an object (its appearance) and their knowledge about it (its real identity), and thus as a way of testing their ability to attribute mental states to themselves (Baron-Cohen, 1989, p. 591). Autistic children fail on this task, and this suggests, according to Baron-Cohen, that they lack awareness of their own mental states (Baron-Cohen, 1989, p. 596). But how is this argument actually supposed to work, and why should an inability to succeed on the appearance-reality task entail a lack of self-awareness? Although it is true that success on the appearance-reality task will call for some measure of access to one's own mental states, any argument to the effect that a person who fails on this task will also lack such access is about as valid as arguing that since writing one's autobiography involves the use of self-awareness, a person who is unable to write his memoirs will also lack self-awareness. The argument is invalid, since it confuses necessary and sufficient conditions. Even if success on the appearance-reality task justifies ascribing self-awareness to the person in question, one cannot conclude that somebody who fails on the task will lack self-awareness. Such a conclusion would only be warranted if the ability to distinguish appearance and reality were a necessary requirement for possessing self-awareness, and no arguments have been put forth to show that that should be the case. Furthermore, as Nichols and Stich have pointed out, even if the argument was valid, the experimental data do not at all support the conclusion drawn by Baron-Cohen, but rather the exact opposite one. As the example with the fake chocolates shows, autistic children have a tendency to fail on the appearance-reality task by being taken in by the *appearance* of the object, i.e., by the way they *experience* the object. Had they really had no access to their own mental states, this is not the kind of error they would have been expected to make. Rather, they should have been very good at detecting what the object really is, and should have had no access to its appearance (Nichols & Stich, 2002, pp. 26–7).

3 Frith and Happé

In their article 'Theory of mind and self-consciousness: What it is like to be autistic?' Frith and Happé argue that autism involves a kind of mind-blindness and that autists can only judge their own mental states by their actions. They also write that the inability to attribute mental states to self is the same as not having introspective awareness (Frith and Happé, 1999, p. 1), and they suggest that an autistic mind might only contain first-order representations of events and experiences (Frith and Happé, 1999, p. 8). As they stress, it is not that autists lack mental states; it is only that they are unable to *reflect* upon their mental states. Autists simply lack the cognitive machinery to represent their thoughts and feelings *as* thoughts and feelings (Frith and Happé, 1999, p. 7); that is, they might not be able to conceptualize their own intentions as intentions (Frith and Happé, 1999, p. 9). These descriptions seem at first to point in the same direction as some of Baron-Cohen's remarks, namely that autism might involve a defect in higher cognition, might involve defective conceptual skills. In the end, however, their article contains an ambiguity related to the one found in Baron-Cohen's contribution, since Frith and Happé do not only speak of autism as involving an *impaired* self-consciousness, but also repeatedly write that autism involves a *lack* of self-consciousness (Frith and Happé, 1999, p. 8). And, of course, to have impaired self-consciousness and to lack self-consciousness are two very different things. Is their claim that autists have mental states, but that they lack self-awareness in the sense of an experiential (immediate and non-inferential) access to these states, or is the claim rather that autists have mental states, that these states (perceptions, desires, beliefs, emotions) feel like something to the autists, that there are experiences involved, but that the autists simply have difficulties articulating these experiences reflectively and conceptually? The situation is not clarified by the fact that Frith and Happé do not offer an account of what exactly they mean by the term 'self-consciousness'.

IV: The Contribution of Phenomenology

It is relatively uncontroversial that we are in possession of theoretical knowledge about other persons and about the workings of our own minds, but the crucial question is whether this theoretical knowledge constitutes the whole of what we call upon when we understand others and ourselves.[6]

1. Self-awareness and the first-person perspective

What has phenomenology had to say about our access to our own mental states, what does this tradition have to say about self-awareness? Needless to say, phenomenologists would insist that we do have a non-inferential access to our own experiential dimension, and that any investigation of consciousness that ignored this first-personal access would be seriously flawed. Moreover, phenomenologists have typically argued that self-awareness cannot be reduced to reflective (thematic, conceptual, mediated) self-awareness. On the contrary, reflective self-awareness or self-knowledge presupposes pre-reflective (un-thematic, tacit, non-linguistic, immediate) self-awareness (cf. Zahavi & Parnas, 1998; Zahavi, 1999).

In contrast to what is claimed by the theory-theory, self-awareness is not something that only comes about the moment I construct a theory about the cause of my own behaviour, a theory that postulates the existence of mental states. On the contrary, it is legitimate to speak of a primitive type of self-awareness whenever I am conscious of an experience — be it a feeling of joy or fatigue or a perception of a sunflower — from a first-person perspective. If the experience is given in a first-personal mode of presentation to me, it is (at least tacitly) *given* as *my* experience, and therefore counts as a case of self-awareness.

Consciousness is not only given when we reflect upon it, it is already given prior to reflection, namely whenever we undergo an experience in its first-personal mode of givenness; that is, whenever there is a 'what it is like' involved with its inherent quality of 'mineness'. This is why Sartre can write that it is just as necessary for an experience to exist self-consciously, as it is for an extended object to exist three-dimensionally (Sartre, 1976, pp. 20–1, Sartre, 1948, pp. 64–5, cf. Husserl, 1987, p. 89).

Given this definition, it should be obvious that the phenomenological discussion of self-awareness has affinities to the contemporary discussion of phenomenal consciousness. In fact, phenomenal consciousness is precisely interpreted as

---

[6] As Heal has pointed out, our theoretical knowledge is much too general and abstract to cover everything. Only a small part of our 'know how' can be reflected and articulated in a theoretical 'know that', and by arguing the way they do, the theory-theorists are basically facing the same problem that research in Artificial Intelligence has also faced and failed to solve, namely the so-called frame problem, i.e., the problem of providing a general theory of relevance (Heal, 1996, pp. 78, 81; cf. Dreyfus, 1992).

a primitive form of self-awareness. To undergo an experience necessarily means that there is something 'it is like' for the subject to have that experience. But insofar as there is something it is like for the subject to have the experience, there must be some awareness of the experience itself; in short, there must be some minimal form of self-awareness (cf. Zahavi, 1999; 2002). To be in possession of this type of self-awareness is consequently simply to be conscious of one's occurrent experience. The self-awareness in question does not exist apart from the experience, as an additional mental act. Rather, it is an intrinsic feature of the experience, and is not brought about by some kind of reflection or introspection.[7]

It is important to emphasize that self-awareness is not an unequivocal concept. To recognize the existence of a primitive form of pre-reflective self-awareness is not to deny the existence of more elaborate forms of self-awareness that might be both theory- and language-dependent and intersubjectively constituted.

2 Primary intersubjectivity

According to the theory-theory, our understanding of other people is in principle like our understanding of stars, clouds, and geological formations. Other people are just complex objects in our environment whose behaviour we attempt to predict and explain, but whose causal innards are hidden from us (Heal, 1986, p. 135), which is part of the reason why we have to make appeal to theory and postulate 'unobservables' such a beliefs and desires. But couldn't it be argued that our understanding of other people differs in fundamental ways from our understanding of inanimate objects? Couldn't it be argued that other people are subjects like ourselves, and that this makes the epistemic situation completely different?

To put it differently, the crucial question is not whether we can predict and explain the behaviour of others, and if so, how that happens, but rather whether such prediction and explanation constitute the primary and most fundamental form of intersubjectivity. As Heidegger has famously argued, the very attempt to grasp the mental states of others is the exception rather than the rule. Under normal circumstances, we understand each other well enough through our shared engagement in the common world (Heidegger, 1979, pp. 334–5), and it is only if this understanding for some reason breaks down, for instance if the other behaves in an unexpected and puzzling way, that other options kick in and take over, be it inferential reasoning or some kind of simulation. By conforming to shared norms, much of the work of understanding one another doesn't really have to be done by us. The work is already accomplished (McGeer, 2001, p. 119).

According to the theory-theory of mind, my access to the mind of another is always inferential in nature, is always mediated by his bodily behaviour. This way of raising and tackling the problem of intersubjectivity has been criticized by phenomenologists. The criticism has been multifaceted, but let us mention

---

[7] This phenomenological approach to self-awareness is consequently in clear opposition to both the so-called reflection model of self-awareness, as well as to the currently so popular higher-order representation theories of consciousness. For an explicit criticism of the latter, cf. Zahavi & Parnas 1998; Zahavi 1999; 2002.

one influential argument.[8] It has been argued that such an account presupposes a highly problematic dichotomy between inner and outer, between experience and behaviour. For phenomenologists, a solution to the problem of other minds must start with a correct understanding of the relation between mind and body. But if we begin with a radical division between a perceived body and an inferred mind we might never, to use Hobson's memorable phrase, be able to 'put Humpty Dumpty together again' (Hobson, 1993, p. 104). Is it really true that experiences and mental states are hidden, unobservable, theoretically postulated entities? When it comes to my own experiential states, such a view seems highly counterintuitive, to put it mildly. But even in the case of others' experiences, the assumption is questionable. When I see another's face, I *see* it as friendly or angry, etc., that is, the very face expresses these emotions. In some sense, experiences are not internal, not hidden in the head, but already present in bodily gestures and actions. As both Merleau-Ponty and Scheler have argued, the affective and emotional experiences of others are given for us *in* expressive phenomena. Anger, shame, hate and love are not only qualities of subjective experience, but also types of behaviour or styles of conduct, which are visible from the outside. They exist *on* this face or *in* those gestures, not hidden behind them (Merleau-Ponty, 1964, pp. 52–3; Scheler, 1973, p. 254). Moreover, bodily behaviour is meaningful, it is intentional, and as such, it is neither internal nor external, but rather beyond this abstract and artificial distinction. When perceiving the actions and expressive movements of other persons, one sees them as meaningful and goal-directed. No inference to a hidden set of mental states is required. Based on considerations like these, phenomenologists have argued that we do not first perceive a physical body in order then to infer, in a subsequent move, the existence of a foreign subjectivity. In other words, intersubjective understanding is not a two-stage process of which the first stage is the perception of meaningless behaviour, and the second is an intellectually based attribution of psychological meaning. On the contrary, in the face-to-face encounter, we are neither confronted with a mere body, nor with a hidden psyche, but with a unified whole. We see the anger of the other, we empathize with his sorrow, we comprehend his linguistically articulated beliefs, we do not have to *infer* their existence. Thus, in many situations, we have a direct, pragmatic understanding of the intentions of others. When we seek to understand others, we do not normally, and at first, attempt to classify their actions under lawlike generalizations, rather we seek to make sense of them. And the question is whether this process of sense-making is theoretical in nature, or whether it rather involves some kind of embodied (emotional and perceptual) skill or practice (Gallagher, 2001, p. 85).

In his book *Mindblindness*, Baron-Cohen introduces three mechanisms that he considers to be precursors to a theory of mind. First, there is what he calls an *Intentionality Detector* . This detector is a device that interprets motion stimuli in terms of primitive volitional mental states. It is very basic. It works through the senses and provides the child with a direct and non-theoretical understanding of

---

[8] For a more extensive discussion and presentation of phenomenological theories of intersubjectivity, cf. Zahavi, 2001a and 2001b.

the intentionality of the other (Baron-Cohen, 1995, pp. 32–4). Secondly, there is what Baron-Cohen calls an *Eye-direction Detector* . This device works through vision, it detects the direction of the eyes of others, and interprets gaze as seeing (Baron-Cohen, 1995, p. 39). Finally, there is the *Shared-attention Mechanism* that enables the child to become aware that another is attending the same object as itself (Baron-Cohen, 1995, pp. 44–5).

According to Baron-Cohen, these three mechanisms, which all become available in the first year of life, enable the child to understand two crucial intentional properties, namely aboutness and aspectuality (Baron-Cohen, 1995, p. 56). What is remarkable, of course, is that even Baron-Cohen, a prominent defender of the theory-theory of mind, concedes that children are in possession of a non-theoretical understanding of the intentions of others.[9] The question then is whether these basic forms of intersubjectivity constitute a constant operative basis, or whether they rather, as Baron-Cohen claims, are infancy precursors to a theory of mind (Baron-Cohen, 2000, p. 1251). In our view, Gallagher is exactly right when he — articulating an insight to be found among many of the classical phenomenologists — writes as follows,

> Primary, embodied intersubjectivity is not primary simply in developmental terms. Rather it remains primary across all face-to-face intersubjective experiences, and it subtends the occasional and secondary intersubjective practices of explaining or predicting what other people believe, desire or intend in the practice of their own minds (Gallagher, 2001, p. 91).

3 Phenomenology and autism

Given this quick overview of what a phenomenological understanding of intersubjectivity and self-awareness amounts to, can phenomenology contribute to a better understanding of autism?

The experimental data presented by the theory-theorists are compatible with the view that autists are in possession of pre-reflective self-awareness — they do experience perceptions, desires, beliefs and emotions from a first-person perspective. This fact, however, does not rule out that many autists might simultaneously suffer from a basic self-estrangement, nor does it preclude many autists from having an impaired propositional self-knowledge as a result of widespread deficits in their cognitive and linguistic capabilities. To put it differently, and this is exactly a suggestion that phenomenology has the conceptual resources to develop further, the findings presented by the theory-theorists are compatible

---

[9] There is plenty of developmental evidence that even in early infancy the young child is capable of tracking others' intentions. This can, for instance, be seen in the phenomena of teasing and in social referencing. To quote from Rochat's recent work: 'Research shows that by nine months infants begin to treat and understand others as "intentional agents", somehow explicitly recognizing that like themselves, people plan and are deliberate in their actions. So, for example, infants will start sharing their attention toward objects with others, looking up toward them to check if they are equally engaged. They will start to refer to other people socially, and in particular to take into considerations the emotional expression of others while planning actions or trying to understand a novel situation in the environment. They will, for instance, hesitate in crossing the deep side of a visual cliff if their mothers express fear [. . .]' (Rochat, 2001, p. 185).

with the view that although autists are *self-aware* in a primitive or primordial sense, many might have difficulties achieving a more robust and comprehensive *self-knowledge* (cf. Raffman, 1999).[10]

We have seen that the theory-theorists emphasize the inferential, theory-driven nature of the intersubjective encounter. One additional way to assess the plausibility of this conjecture is to look for first-person accounts, where autists describe their own experience of interpersonal relations. In her conversations with Oliver Sacks, Temple Grandin provides an illuminating account:

> She was at pains to keep her own life simple, she said, and to make everything very clear and explicit. She had built up a vast library of experiences over the years, she went on. They were like a library of videotapes, which she could play in her mind and inspect at any time — 'videos' of how people behaved in different circumstances. She would play these over and over again and learn, by degrees, to correlate what she saw, so that she could then predict how people in similar circumstances might act. She had complemented her experience by constant reading, including reading of trade journals and the *Wall Street Journal* — all of which enlarged her knowledge of the species. 'It is strictly a logical process,' she explained. [...] When she was younger, she was hardly able to interpret even the simplest expressions of emotion; she learned to 'decode' them later, without necessarily feeling them. [...] What is it, then, I pressed her further, that goes on between normal people, from which she feels herself excluded? It has to do, she has inferred, with an implicit knowledge of social conventions and codes, of cultural presuppositions of every sort. This implicit knowledge, which every normal person accumulates and generates throughout life on the basis of experience and encounters with others, Temple seems to be largely devoid of. Lacking it, she has instead to 'compute' others' intentions and states of mind, to try to make algorithmic, explicit, what for the rest of us is second nature. She herself, she infers, may never have had the normal social experiences from which a normal social knowledge is constructed (Sacks, 1995, pp. 248, 257–8).

What we are confronted with here is not a lack of a theory of mind, but with a lack of an immediate, pre-reflective, or implicit understanding of the meaning of social interaction. Occasionally, this deficit is compensated by an intellectual, theory-driven approach. But this clinical picture is straightforwardly in contradiction with the claims made by the proponents of the theory-theory. In fact, it seems to us that Grandin's compensatory way of understanding others perfectly resembles how *normal* intersubjective understanding is portrayed by the

---

[10] It is worth noticing that Nichols and Stich have recently argued that mind-reading skills should be divided into the two categories: detecting and reasoning. Roughly put, the idea is that different mechanisms are active in detecting (attributing, having access to) mental states and reasoning about (explaining) them (Nichols & Stich, 2002, p. 2). This distinction appears very reasonable, and so does a natural inference, namely that only the process of reasoning involves a theory of mind. However, Nichols and Stich also argue that in order to be in possession of self-awareness all that is needed is a monitoring mechanism that 'takes a representation p in the Belief Box as input and produces the representation *I believe that p* as output'. Thus, all the 'Monitoring Mechanism has to do is copy representations in the Belief Box, embed them in a representation schema of the form: *I believe that . . .* , and then place this new representation back in the Belief Box'(Nichols & Stich, 2002, pp. 13, 40). Whether this attempt to account for the structure of self-awareness and solve the classical problems concerning *first-person reference* is successful, is, however, another question, and in our view highly unlikely.

proponents of the theory-theory. To put it somewhat ironically, the evidence suggests that autists (provided that they posses a sufficiently high IQ) might be more characterized by an excessive reliance on a theory of mind (in the proper sense of the word) than by a lack of such a theory. They seem to have to rely on wooden algorithms and formulas if they are to understand other people. This is not to deny, of course, that autists have difficulties passing theory of mind tasks. Rather, the point is that they have these difficulties not because of a lack of a theory of mind, but because of other deficiencies. In this sense, it really is question-begging to label false-belief tasks or appearance-reality tasks as 'theory of mind' tasks; it prejudges the issue by suggesting that psychological competence consists in the possession and use of a theory.[11]

There is long tradition of applying phenomenological tools to psychiatry, and although few phenomenologists have worked explicitly with infantile autism,[12] many have worked on schizophrenic autism. Some of the insights obtained in this work are also of relevance for the current discussion.

Thus, ideas developed in the theory-theory approach to infantile autism have also been applied to schizophrenia;[13] briefly put, schizophrenia has been considered a 'late-onset autism' (Frith, 1994, p. 150), and schizophrenic deficiency has been understood as 'an inability to represent our own mental states, including our intentions' (Frith 1994, p. 151). Chris Frith has specifically proposed that schizophrenics suffer a deficit in 'theory of mind'. 'Schizophrenic patients', Frith writes, 'lack *awareness of their own mental states*, as well as the mental states of other people'(Frith, 1994, p. 151, our italics). Again, it is not quite clear what Frith is referring to when he talks of a lack of awareness — whether it is a lack of direct access to own mental states such as pains, desires and beliefs, or whether it involves a diminished ability to form higher-order beliefs about beliefs. But in general, empirical literature does not provide evidence that schizophrenics fail on theory of mind tests in the same way as infantile autists do (in schizophrenia, diminished performance on theory of mind tests is a state phenomenon, i.e., only

---

[11] As mentioned before, autism is generally taken to corroborate the existence of a theory of mind module. The selective impairment on mind-reading tasks has been explained by reference to a damaged or destroyed theory of mind mechanism. But as Boucher has pointed out, this single primary deficit theory faces some obvious objections. Infants and many learning-disabled people do not have a theory of mind, but they are not autistic (in fact, children with Down's syndrome who are noticeable deficient in theoretical abilities do have the psychological competence necessary to understand others) and people with Asperger's syndrome — so-called high-functioning autists — are autistic but do, in fact, have a theory of mind, and succeed well when it comes to false-belief tasks (Boucher, 1996, p. 233). The lack of a theory of mind consequently seems to constitute neither a sufficient nor a necessary condition for autism. This also suggests that the psychological deficits underlying autism cannot be in the area of higher cognitive functions. This suggestion is further corroborated by the fact that the first signs of autism already start to appear during the first year, i.e., long before the supposed appearance of a theory of mind. These signs include the marked absence of any form of joint-attention behaviour. In recent years, this has made mind-reading theories increasingly identify the precursors of a theory of mind, rather than the lack of theory of mind itself, as the primary cause of autism (cf. Baron-Cohen, 2000, p. 1251).

[12] Bosch (1970) is a rare exception.

[13] Note that the term 'autism' was first introduced in schizophrenia by Bleuler, and signified a retreat to an inner world, blooming with mentalizing activity (Bleuler, 1911).

occurring during psychotic exacerbations, and is mainly restricted to patients with disorganized schizophrenia [Parnas & Sass, 2001]). From a clinical point of view, we are rarely faced with a universal absence of, or even diminishment of, 'mentalizing' or self-consciousness in schizophrenia (Sass & Parnas, 2003): consider, for example, that the world of the paranoid schizophrenic may well be bristling with complex and malevolent, mental or intentional states, often experienced as being directed toward the patient; and that, as Blankenburg, Conrad, Sass and many other authors have aptly demonstrated, schizophrenic persons often demonstrate an exaggerated and all-encompassing kind of self-consciousness with rich proliferation of metalevels (Minkowski, 1927; Blankenburg, 1971; Conrad, 1958; Sass, 1992); a proliferation which appears as compensatory to the more basic deficits in the immediate, pre-reflective understanding of the world and the others. More specifically, it appears that interpersonal understanding in schizophrenia involves a general diminishment of a direct, tacit and pre-reflective grasp of meaning (a state often described as perplexity, cf. Berze, 1914; Störing, 1939), and a compensatory reliance on reflective inferential reasoning, a cognitive schizophrenic style traditionally labelled as 'morbid rationalism' (cf. Minkowski, 1927; Parnas et al., 2002). Consider the following complaints about diminished capacity for direct grasp of others and world:

> What is it that I am missing? It is something so small, but strange, it is something so important. It is impossible to live without it. I find that I no longer have footing in the world. I have lost a hold in regard to the simplest, everyday things. It seems that I lack a natural understanding for what is a matter of course and obvious to others.... I am missing the basics.... I don't know how to call this.... It is not knowledge.... Every child knows it!! (Schizophrenic patient reported in Blankenburg, 2001, pp. 307–8).

Recent, empirical studies of patients with schizophrenia spectrum disorders confirm that the most significant aspects (especially of the early stages of the illness) comprise a loss of the immediate, pre-conceptual attunement to others and the world, and that this is frequently associated with profound alterations in the pre-reflective sense of self (Parnas & Handest, 2003; Parnas et al., 2003).

In summary and conclusion: we think that phenomenology may be as helpful for cognitive science as it appears to be for psychopathology (cf. Parnas & Zahavi, 2002). First, it can assist scientists in developing a more rigorous and clarified use of mental terms, and it can provide a conceptual framework suitable for addressing the structures and processes of subjectivity, a framework that transcends a simple introspective methodology, and which can inform the design of the experimental study. Thus, one could, for instance, envisage setting up a study specifically focussed on comparing performance on tasks involving intuitive-direct understanding and tasks relying on inferential reasoning.

Acknowledgments

This study was funded by the Danish National Research Foundation. Thanks to Shaun Gallagher for comments to an earlier version of the article.

References

Avramides, A. (2001), *Other Minds* (London: Routledge).
Baron-Cohen, S. (1989), 'Are autistic children "behaviorists"? An Examination of Their Mental–Physical and Appearance–Reality Distinctions,' *Journal of Autism and Developmental Disorders*, **19** (4), pp. 579–600.
Baron-Cohen, S., Leslie, A., Frith, U. (1985), 'Does the autistic child have a "theory of mind"?', *Cognition*, **21**, pp. 37–46.
Baron-Cohen, S. (1995), *Mindblindness: A Essay on Autism and Theory of Mind* (Cambridge, MA: MIT Press).
Baron-Cohen, S. (2000), 'The cognitive neuroscience of autism: Evolutionary approaches', in *The New Cognitive Neurosciences* (second edition), ed. M.S. Gazzaniga (Cambridge, MA: MIT Press).
Berze, J. (1914), *Die Primäre Insuffizienz der Psychischen Aktivität Ihr Wesen, ihre Erscheinungen und ihre Bedeutung als Grundstörungen der Dementia Praexox und den hypophrenen Überhaupt* (Leipzig: F. Deuticke).
Blankenburg, W. (1971), *Der Verlust der Natürlichen Selbstverständlichkeit: Ein Beitrag zur Psychopathologie Symptomarmer Schizophrenien* (Stuttgart: Ferdinand Enke Verlag).
Blankenburg, W. (2001), 'First steps toward a psychopathology of "common sense"', *Philosophy, Psychiatry, Psychology*, **8**, pp. 303–15.
Bleuler, E. (1911/1950), *Dementia Praecox or the Group of Schizophrenias*, trans. J. Zinkin (New York: International Universities Press).
Bosch, G. (1962/1970), *Infantile Autism: A Clinical and Phenomenological-Anthropological Approach Taking Language as the Guide* (New York: Springer-Verlag).
Botterill, G. (1996), 'Folk psychology and theoretical Status', in *Theories of Theories of Mind*, ed. P. Carruthers & P.K. Smith (Cambridge: Cambridge University Press).
Boucher, J. (1996), 'What could possibly explain autism?', in *Theories of Theories of Mind*, ed. P. Carruthers & P.K. Smith (Cambridge: Cambridge University Press).
Carruthers, P. (1996a), 'Simulation and self-knowledge: a defence of theory-theory', in *Theories of Theories of Mind*, ed. P. Carruthers & P.K. Smith (Cambridge: Cambridge University Press).
Carruthers, P. (1996b), 'Autism as mind-blindness: an elaboration and partial defence', in *Theories of Theories of Mind*, ed. P. Carruthers & P.K. Smith (Cambridge: Cambridge University Press).
Carruthers, P., Smith, P. K. (1996), 'Introduction', in *Theories of Theories of Mind*, ed. P. Carruthers & P.K. Smith (Cambridge: Cambridge University Press).
Conrad, K. (1958), *Die beginnende Schizophrenie. Versuch einer Gestaltanalyse des Wahns* (Stuttgart: Thieme Verlag).
Dreyfus, H.L. (1992), *What Computers Still Can't Do: A Critique of Artificial Reason* (Cambridge, MA: MIT Press).
Frith, U., Happé, F. (1999), 'Theory of Mind and Self-Consciousness: What Is It Like to Be Autistic?', *Mind & Language*, **14** (1), pp. 1–22.
Frith, C.D. (1994), 'Theory of mind in schizophrenia', in *The Neuropsychology of Schizophrenia*, ed. A. David, J. Cutting (Hillsdale, NJ: Erlbaum).
Gallagher, S. (2001), 'The Practice of Mind: Theory, Simulation or Interaction', *Journal of Consciousness Studies* **8** (5–7), pp. 83–108.
Gopnik, A. (1993), 'How we know our minds: The illusion of first-person knowledge of intentionality', *Behavioral and Brain Sciences*, **16**, pp. 1–14.
Gopnik, A. (1996), 'Theories and modules: creation myths, developmental realities, and Neurath's boat', in *Theories of Theories of Mind*, ed. P. Carruthers & P.K. Smith (Cambridge: Cambridge University Press).
Gopnik, A., Meltzoff, A.N., Kuhl, P.K. (2001), *The Scientist in the Crib. What early learning tells us about the mind* (New York: Perennial).
Gopnik, A., Wellman, H.M. (1995), 'Why the child's theory of mind really *is* a theory', in *Folk Psychology: The Theory of Mind Debate*, ed. M. Davies & T. Stone (Oxford: Blackwell).
Gordon, R.M. (1996), '"Radical" simulationism', in *Theories of Theories of Mind*, ed. P. Carruthers & P.K. Smith (Cambridge: Cambridge University Press).
Gurwitsch, A. (1966), *Studies in Phenomenology and Psychology* (Evanston: Northwestern University Press).
Heal, J. (1996), 'Simulation, theory, and content', in *Theories of Theories of Mind*, ed. P. Carruthers & P.K. Smith (Cambridge: Cambridge University Press).
Heidegger, M. (1979), *Prolegomena zur Geschichte des Zeitbegriffs* (Frankfurt am Main: Vittorio Klostermann).
Heidegger, M. (1993), *Grundprobleme der Phänomenologie (1919)* (Frankfurt am Main: Vittorio Klostermann).
Hobson, R.P. (1993), *Autism and the Development of Mind* (Hove: Psychology Press).
Husserl, E. (1984), *Einleitung in die Logik und Erkenntnistheorie* (Dordrecht: Martinus Nijhoff).
Husserl, E. (1987), *Aufsätze und Vorträge (1911–1921)* (Dordrecht: Martinus Nijhoff).

McGeer, V. (2001), 'Psycho-practice, Psycho-theory and the Contrastive Case of Autism', *Journal of Consciousness Studies* **8** (5–7), pp. 109–32.
Merleau-Ponty, M. (1964), *Sense and Non-Sense* (Evanston, IL: Northwestern University Press).
Minkowski, E. (1927), *La schizophrénie* (Paris: Payot & Rivages).
Nichols, S., Stich, S. (2002), 'Reading One's Own Mind: A Cognitive Theory of Self-Awareness', http://ruccs.rutgers.edu/ArchiveFolder/Research%20Group/Publications/Room/ room.html.
Parnas, J., Sass, L.A. (2001), 'Self, solipsism, and schizophrenic delusions', *Philosophy, Psychiatry, Psychology*, **8**, pp. 101–20.
Parnas, J., Zahavi, D. (2002), 'The role of phenomenology in psychiatric diagnosis and classification', in *Psychiatric Diagnosis and Classification*, ed. M. Maj *et al.* (Chichester: John Wiley & Sons).
Parnas, J., Bovet, P., Zahavi, D. (2002), 'Schizophrenic autism: Clinical phenomenology and pathogenetic implications', *World Psychiatry*, **1** (3), pp. 131–6.
Parnas, J., Handest, P. (2003), 'Phenomenology of anomalous self-experience in early schizophrenia', *Comprehensive Psychiatry*, **44** (2), pp. 121–34.
Parnas, J., Handest, P., Sæbye, D., Jansson, L. (2003), 'Anomalies of subjective experience in schizophrenia and psychotic bipolar illness', *Acta Psychiatrica Scandinavica*, **118**, pp. 126–33.
Raffman, D. (1999), 'What autism may tell us about self-awareness: A commentary on Frith and Happé', *Mind & Language* **14** (1), pp. 23–31.
Rochat, P. (2001), *The Infant's World* (Cambridge, MA: Harvard University Press).
Sacks, O. (1995), *An Anthropologist on Mars* (London: Picador).
Sartre, J.-P. (1943/1976), *L'être et le néant* (Paris: Gallimard).
Sartre, J.-P. (1948), 'Conscience de soi et connaissance de soi,' *Bulletin de la Société Française de Philosophie*, **XLII**, pp. 49–91.
Sass, L. (1992), *Madness and Modernism* (New York: Basic Books).
Sass, L., Parnas, J. (2003), 'Schizophrenia, consciousness and the self', *Schizophrenia Bulletin*, in press.
Scheler, M. (1973), *Wesen und Formen der Sympathie* (Bern/München: Francke Verlag).
Störring, G. (1939/1987). 'Perplexity', in *The Clinical Roots of the Schizophrenia Concept*, ed. J. Cutting & M. Shepherd (Cambridge, UK: Cambridge University Press), pp 79–82.
Zahavi, D., Parnas, J. (1998), 'Phenomenal consciousness and self-awareness. A phenomenological critique of representational theory', *Journal of Consciousness Studies* **5** (5–6), pp. 687–705.
Zahavi, D. (1999), *Self-awareness and Alterity: A Phenomenological Investigation* (Evanston, IL: Northwestern University Press).
Zahavi, D. (2001a), 'Beyond empathy: Phenomenological approaches to intersubjectivity', *Journal of Consciousness Studies*, **8** (5–7), pp. 151–78.
Zahavi, D. (2001b), *Husserl and Transcendental Intersubjectivity* (Athens: OH: University Press).
Zahavi, D. (2002), 'First-person thoughts and embodied self-awareness. Some reflections on the relation between recent analytical philosophy and phenomenology', *Phenomenology and the Cognitive Sciences*, **1**, pp. 7–26.
Zahavi, D. (2003), *Husserl's Phenomenology* (Stanford: Stanford University Press).

Patrick Haggard and Helen Johnson
Experiences of Voluntary Action

Psychologists have traditionally approached phenomenology by describing perceptual states, typically in the context of vision. The control of actions has often been described as 'automatic', and therefore lacking any specific phenomenology worth studying. This article will begin by reviewing some historical attempts to investigate the phenomenology of action. This review leads to the conclusion that, while movement of the body itself need not produce a vivid conscious experience, the neural process of voluntary action as a whole has distinctive phenomenological consequences.

The remainder of the article tries to characterise this phenomenology. First, the planning of actions is often conscious, and can produce a characteristic executive mode of awareness. Second, our awareness of action often arises from the process of matching what we intended to do with what actually happened. Failures of this matching process lead to particularly vivid conscious experience, which we call 'error awareness'. These features of action phenomenology can be directly related to established models of motor control. This allows an important connection between phenomenology and neuroscience of action. Third, whereas perceptual phenomenology is normally seen as caused or driven by the sensory stimulus, a much more fluid model is required for phenomenology of action. Several experimental results suggest that phenomenology of action is partly a post hoc reconstruction, while others suggest that our awareness of action represents an integration of several processes at multiple levels of motor processing. Fourth, and finally, studies of the phenomenology of action, unlike those of perception, show a strong linkage between primary awareness and secondary awareness or self-consciousness: awareness of action is specifically and inextricably awareness of my action. We argue that the concepts of agency and of proprioaction (my control over my own body) are fundamental to this linkage. For these reasons, action represents a much more promising field than perception for attacking the problematic question of the relation between primary and secondary consciousness. Some promising directions for future research are indicated.

Introduction

This article offers the perspective of a psychologist on the conscious experience of action. An action is defined as a movement of the body, resulting from specific mental preparation, and aimed at some goal that the agent desires to achieve. It is a neglected perspective, in the sense that relatively few psychologists work on or are trained in the psychology of action, and few of those consider the phenomenology associated with controlling actions. This is a result of the place that action holds in the two frameworks that dominated psychology in the latter part of the twentieth century: behaviourism and cognitive psychology.

Behaviourism contributed little to our understanding of the psychological processes of action. This seems in one way surprising, since behaviourism famously insisted on overt motor acts being the only permissible object of psychological investigation (Skinner, 1957). However, behaviourism typically used motor actions as a marker of stimulus processing, rather than as a specific process in their own right. In addition, the behaviourists' reliance on animal conditioning as a primary source of data limits both the concept of voluntary action, and the study of phenomenology. Cognitive psychology was similarly restrictive, though for different reasons. Cognitive paradigms focussed on symbolic operations on representations (Fodor, 1981), drawing their ultimate inspiration from language (Fodor, 1975). The core computational problems of action, in contrast, focus on how a neural and mental representation of a goal leads to a coordinated physical movement of the body. Symbolic representation does not figure in this question. For these reasons, perhaps, the control of action has not been a central topic in psychology in the twentieth century.

Thus, psychologists working on the neural aspects of action, or motor control, may have viewed their work as neuroscience, as much as psychology. A corollary of neuralising the psychology of action has been a lack of trust in the subject. Goal-directed action generally involves both a neural process and a conscious experience. However, this latter aspect, the phenomenology of action, has been unjustly neglected. Why might this be? The most obvious explanation is that the phenomenology of action is often thin: we perform many actions without vivid conscious experience of what we are doing. Our actions are often automatic. Perhaps psychologists have taken automaticity to mean that actions have no phenomenology at all. In fact, the phenomenology of action normally remains background, but it is essential to the way we act, and it but can be very vivid under appropriate circumstances, as will be discussed later.

Introspective Phenomenology

Looking further back to the birth of psychology as a science in the late nineteenth and early twentieth centuries, it becomes clear that this neglect is relatively new. In reviewing one hundred years of British psychology of action (Haggard, 2001), we were struck by the complex and structured attempts made by early psychologists to understand the conscious experience of action. This work was developed by Ward and Stout, and often relied as much on philosophical conceptual analysis as on experimental data. The key concept in this tradition was *conation*. This corresponds roughly to our concept of intentional action (Searle, 1983); a conscious mental state aiming to produce a specific goal through a physical action. A more detailed review of the concept can be found elsewhere (Haggard, 2001). Here, the key point is that introspection was taken as an important part of psychological method. Stout (1906), for example, clearly trusted the subject

> What we have called 'felt tendency' (i.e. conation) seems to elude introspection as carried out under the test conditions of the laboratory.... Those psychologists (who are distinctly adverse to the existence of any peculiar kind of immediate experience distinctively characteristic of conation)...do not look for [conation] where alone it can be found, and their failure to find it is therefore not a sufficient reason for denying its existence (p.11).

In this paper, we review a second tradition in the psychological phenomenology of action. We have already seen that the British psychologists' approach to the psychology of action was based on conceptual analysis. Indeed, its lasting influence has been in the philosophy of action, rather than in psychology (e.g., Anscombe, 1963; Candlish, 1981). At around the same time, the continental tradition was developing an experimental approach to the psychology of action, which was also explicitly phenomenological. This tradition emerged from the focus of the Wurzburg school on experimental studies of thought and higher cognitive function (Külpe, 1893). The Wurzburg tradition was extended to voluntary actions by Narziss Ach (1905). Ach's method consisted of giving the subjects what would now be called an 'action script', and asking them to report the succession of experience as they enacted the script. While Ach clearly 'trusted the subject' it is hard to draw general conclusions from his work, because he studied a wide range of actions without clearly isolating or characterising the different elements that jointly comprise the action script. If the action instructions are considered the input to the action system, his work suffers from an unsystematic approach to manipulating the input, and a non-analytic approach.

This deficiency was addressed in a study of Michotte & Prum (1910). This study has apparently been totally ignored by all subsequent work, although it contains by far the most detailed experimental psychological investigation of the conscious experience of action of which we are aware. The core of Michotte and Prum's method was not an action script but simultaneous visual presentation of two numbers. The numbers were preceded by an instruction listing a set of possible arithmetic operators (e.g., multiplication, division). These provided the antecedent context of action. The subject read the numbers, selected voluntarily which arithmetic operation to apply to them, performed the operation, pressed a response key to indicate completion, and then gave a verbal report of their mental states during the process, which included the answer to the sum they had chosen to compute. In this paradigm, freely-selected intentional action occurs through the subject's voluntary decision regarding which arithmetic operation to choose. The most intuitive and ecological consequence of the decision would be to say aloud the number resulting from the arithmetic operation. However, technological limitations in the early days of psychology did not allow accurate recording of verbal reaction times, so the authors substituted a button press as an alternative physical marker of the completion of the mental intention: it can be considered an unusual form of motor output, serving as a proxy outcome of the action. Michotte and Prum understood the action as a whole to include the series of mental processes extending from the presentation of the stimulus materials, the decision regarding the operation to be performed, the process of performing the operation, and the (slightly unusual) physical motor output marking the completion of the process.

The authors' real interest lay in the introspective impression of the various mental experiences during the stream of processing between expecting the numbers, and making the speeded response. The subjective reports revealed a distinct phenomenology of subject's intentions being drawn towards a particular action:

> Mon attention a été attirée, au premier moment, par la juxtaposition du 50, du premier nombre, et du 25 du second: le mot 'division' n'était pas présent, mais je pensais à quelque chose 'qu'il y avait à faire' et ce 'quelque chose' était suffisamment bien déterminé pour que je puisse m'en servir, pour voir si cela pouvait aller ainsi. Je m'aperçus alors tout à coup que cela ne pouvait se faire, de fait (Michotte & Prum, 1910, p. 158)

> (My attention was drawn at first by the juxtaposition of the 50 in the first number and by the 25 in the second. The word 'division' was not yet present in my mind, but I was thinking of something that I had to do, and this 'something' was sufficiently determinate that I could make use of it, to see if it would work out. Suddenly, I noticed that this 'something' really could not be done).

Michotte and Prum's work contains several important messages for the phenomenology of action. First, they show that introspection of action can reveal a reliable and systematic structure of conscious states linked to voluntary action. The key to isolating these states is an experimental method which avoids reducing actions to immediate automatised responses, which allows a true element of free selection, yet which is sufficiently controlled to be conceptually clear and empirically reliable. Free selection is critical to maintain a distinction between voluntary actions and simple reflex actions, which have quite different phenomenologies. Second, they emphasised the importance of the 'tendance de determination' (determining tendency) in voluntary action, as originally introduced by Ach. This term refers to the systematic development of action from a general intention towards the more specific, determinate form of a physical movement. Their subjects experienced the phenomenology of arithmetic actions as a 'push' from a set of numbers, firstly to a decision regarding what to do with them, then to a physical response on realising that operation. Interestingly, two quite different approaches to action also confirm the crucial role of the developing specification of actions. In computational motor control, the motor planning stage transforms a high level description of a goal into a detailed specification of the motor commands required to reach the goal (Wolpert, 1997). In addition, quantitative studies of action awareness concur that the neural process of specifying how an action will be achieved supplies the content of our conscious experience of intentions (Haggard and Eimer, 1999).

Our aim here is not to support or criticise the view of action that Michotte and Prum propose, but to point out how far an introspective phenomenology of action can go. Michotte and Prum used trained subjects (themselves, and their co-workers). But their subjective reports seem to express states which we all experience at least occasionally. Moreover, the reports can be positioned within a well-structured action task, they can be related to an overt behaviour, and are supported by at least some convergent evidence. Michotte and Prum even tried to measure physiological events accompanying voluntary action, and to relate them to conscious experience. They felt that the physiological monitoring altered conscious experience inevitably, and therefore did not pursue the attempt at great length. However, it is interesting to note that they saw the potential of a psychophysiology of intention several decades before Libet (Libet et al., 1983) succeeded with this approach.

The reader of Michotte & Prum (1910) is struck by a paradox. On the one hand, the phenomenology of action is often thin. Psychologists have regularly commented that actions are 'automatic' (Broadbent, 1982). Indeed we generally have minimal conscious experience of our breathing, walking, eating, sports playing etc. This has led one psychologist to suggest we are conscious of very little of what we do (Wegner, 2003). On the other hand, when we wish to, we can report in considerable detail the processes of preparation and execution of our actions, as in Michotte and Prum's experiments. Moreover, when our actions go wrong, the phenomenology is often very strong indeed: many of us will recall, for example, the vivid experience of kicking a penalty which missed the goal and thus lost the match. Our actions may have a strong phenomenology when facilitated by attention. It seems likely that the normal experience of action is a thin and unattended version of this more intense, focal experience. The ability to access a more detailed phenomenology of action may be important in motor learning, rehabilitation and recreational activity.

The continental tradition, exemplified by the painstaking studies of Michotte and Prum (1910) has not lasted. The phenomenology of action has again become an active research topic in recent years, but the feel of this new work is quite different. Whereas the continental tradition relied on the content of systematic introspective report provided by trained observers, the current trend is towards a quantitative psychophysics of action. The quantitative method involves subjects judging a single attribute of their action along a dimension preselected by the experimenter, such as the timing, force or spatial direction of action. It risks missing essential aspects of the phenomenology of action by not sufficiently trusting the subject. In particular, the subject is asked to describe their action only along the dimensions specified by the experimenter. This raises the possibility of asking and answering the wrong question, rather than focussing on the hypothesis under test. In decision theory, this has been called the type III error, to complement the better-known type I and type II errors in statistical inference (Mitroff, 1998). For example, motor control studies have reported that subjects can adjust their movements in response to a change in the location of a visual target of which they are perceptually unaware. This work suggests a dissociation between motor performance and perceptual awareness, but does not address the question of the link between motor performance and motor awareness (see Perceived Spatial Properties of Action section below). No studies of action phenomenology, to our knowledge, have achieved the harmonious combination of rigour of experimental control, depth of introspective report, and power of quantitative psychophysics.

A Framework For Action Awareness

The rest of this paper will focus on the quantitative psychophysics tradition, particularly on studies of the perceived temporal and spatial properties of action. A framework for understanding these studies is shown below (Figure 1).

The framework represents the neural-motor events involved in voluntary action on the left-hand side, and the elements of conscious experience of action on the right-hand side. Thus, making a voluntary action typically involves neural preparation for the action, then the contraction of muscles in response to a motor command, and finally delayed sensory feedback that the body parts have indeed moved. These neural events correspond to the conscious experiences of intention, action and effect respectively, but not in a one-to-one fashion. For example, the conscious experience of intention is driven by a relatively late, motoric stage in neural preparation, namely the point of selecting which specific motor pattern will be used to achieve an action goal (Haggard & Eimer, 1999). Similarly, the perceived time of actions seems to depend, at least in part, on the process of preparing those actions (Haggard *et al.*, 1999). The conscious experience of our actions is an integrated and compressed version of the underlying neural events that cause our actions.

Two elements seem to dominate in this integrated phenomenology of action: the intention to act, and the consequence or effect that the action subsequently

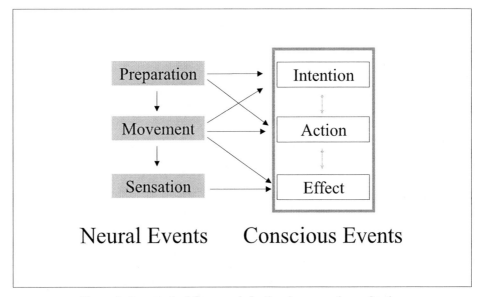

Figure 1. A conceptual framework for the phenomenology of action.

has in the world. In contrast, actual bodily movement is a rather thin centre of the experience of action. For example, our everyday experience tells us that we may really want to eat something. We are aware of hunger, desire to eat, of the taste of food in the mouth, and the subsequent feeling of being full. However, the taste of the food fills our mind much more than the action of chewing it. In fact, as James (1890) and later authors (Bekkering & Wohlschlager, 2002) pointed out, our intentions aim at goals, or effects of our actions in the world. The conscious content of an intention is often simply the goal state that our action will produce. In the above case, our intention to pick up the apple can be described as an attempt to alleviate hunger. Intentions necessarily precede effects, but our experience of intention typically points forwards across this temporal gap, towards their effect.

Perceived Time of Actions

Results of recent studies of the perceived time of events in voluntary action confirm this process of temporal integration in our experience of action. We used the method of Libet *et al.* (1983) to investigate the time at which subjects thought they made a finger movement, or (in separate blocks) the time at which they perceived the onset of an auditory tone (that followed 250 ms after their finger movement). When subjects made the finger movement freely and voluntarily, we observed a shift of the perceived time of the action towards the auditory tone that it caused, compared to a control condition where subjects just made the action without producing the subsequent auditory effect. In addition, subjects perceived the auditory tone as shifted earlier in time towards the action that caused it, in comparison to a control condition where the tone occurred alone without prior voluntary action. The result was that action and effect were perceived as more closely linked in time than they truly were. We coined the term intentional binding to refer to this effect (Haggard *et al.*, 2002). In other conditions, the finger movement was applied involuntarily to the subject, either by moving their finger with a machine (Tsakiris & Haggard, 2003) or by externally stimulating the motor areas of the brain using TMS (Haggard *et al.*, 2002). These involuntary movements reversed the intentional binding effect, producing a repulsion between the perceived time of actions and their consequences. To summarise this work, the presence of an intention to act has a strong structuring influence on the conscious experience of action. This influence takes the form of integration across time, among other phenomena.

We have previously suggested that the phenomenology of action is normally quite thin. However, we become immediately and strongly aware of our actions when the action or its effect is not as intended. We refer to this as 'error awareness'. Indeed, the literature on action slips (Reason, 1990) is one of the few areas of the psychology of action which has recognised phenomenology as well as performance. At the point where an action fails, we often experience a sudden, powerful focus of attention on what we have done, a realisation of the effects of our action, and (most unpleasantly) a realisation that we cannot undo what we have done. Often, time may appear to stand still at that moment:

So saying, her rash hand in evil hour
Forth reaching to the Fruit, she pluck'd, she eat:
Earth felt the wound, and Nature from her seat
Sighing through all her works gave signs of woe,
That all was lost. (Milton, 1667)

Perceived Spatial Properties of Action

Few quantitative psychophysical studies have dealt with error awareness, perhaps because of the difficulty of provoking realistic action errors in laboratory settings. One feasible way of achieving this, however, is the antisaccade or anti-pointing task. In this task the subject makes either a saccadic eye movement, or a reaching movement towards a target. As the subject begins to move, the target is displaced, typically at right angles to the direction of movement. The subject's task is to adjust their movement in the direction opposite to the target shift (Figure 2, right panel).

Thus, if the target jumps to the right during the course of the primary movement, then the subject should adjust their movement of equivalent amplitude, but towards the left. In this situation, subjects frequently follow the target, rather than moving away from it, because they fail to inhibit the prepotent response of saccading or pointing towards a target. Johnson et al. (2002) studied the awareness of the spatial path of pointing movements by asking subjects to make speeded movements which could include either pro-point or anti-point adjustments following an unpredictable shift of the target. Immediately after each

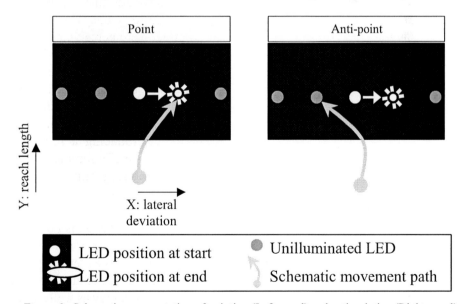

Figure 2. Schematic representation of pointing (Left panel) and anti-pointing (Right panel).

In pointing, the subject reaches for the central target. During movement, the target may jump unpredictably to left or right, and the subject follows the target. During anti-pointing the subject moves in the direction opposite to the target jump. Reproduced from Johnson, van Beers and Haggard (2002).

movement, subjects repeated the spatial path they had just made, without any time constraint. In pro-pointing blocks where subjects followed any shift of the target, subjects underestimated how successfully and rapidly they followed the target shift. In anti-pointing blocks, in contrast, the opposite result was found: subjects overestimated the speed and gain with which they veered away from the displacement of the target. This pattern of results suggests two qualitatively different relations between action and awareness. In normal circumstances (e.g., pro-pointing), the motor system may operate without conscious experience, and much faster than conventional mechanisms of awareness (Castiello et al., 1991). Awareness is a delayed and attenuated version of motor performance. In contrast, tasks such as anti-pointing require an additional mechanism of executive control to inhibit the prepotent response. This executive process constructs a quite different version of awareness, which overestimates the gain and speed of motor performance. Thus, in anti-pointing, Johnson et al.'s subjects seemed to be aware of what they meant to do, rather than of what they actually did.

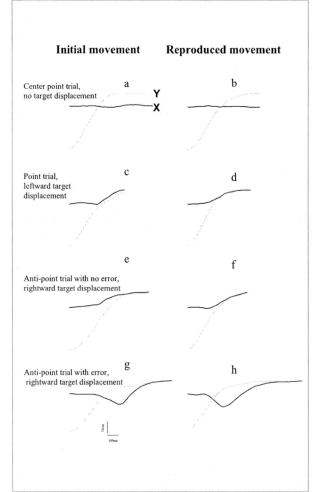

Figure 3. Action awareness for the different kinds of trials studied by Johnson et al. (2002). The thin grey line shows the forward movement of the index fingertip along the Y axis between start and target positions, as a function of time. The solid black line shows the lateral deviation of the index finger in the X axis direction. The visual target might unexpectedly shift along by 5 cm in either direction along the X axis when the subject initated the movement.

After each movement, the subject reproduces the spatial path they have just made: the reproduced movements give an indication of action awareness. Note faithful reproduction in normal pointing (a,b), and underestimation of the speed and amplitude of lateral component of the movement (X dimension, thick black line) following target perturbation in pointing (c,d).

Subjects believed they made anti-point corrections before they actually did (e,f). However, they quite faithfully reproduced the pro-pointing errors that they occasionally made on anti-point trials (g,h), even though the same kinematic features were underestimated in pro-point trials.

Here, however, our interest focuses on the 23% of anti-point trials on which subjects made a pro-pointing error (Figure 3). On these trials, the target might jump to the left, say, as the subject reached towards it. The anti-pointing task requires the subject to deviate to the right, away from the target, by an equal amount. However, on several *error trials*, we were able to detect an initial, incorrect deviation *towards* the target, which was followed by a correct deviation in the opposite direction. As before, we asked subjects to repeat as precisely as possible the spatial path of each movement just after they had made it. Our interest focussed on whether subjects were aware of the initial, erroneous deviation towards the target. The normal pattern of executive construction of awareness in anti-pointing would suggest they should not be. However, results clearly showed that subjects were aware of these initial erroneous deviations. Moreover, the reproduced paths suggested that their awareness of initial erroneous deviations did not show the pattern of delayed, attenuated awareness associated with pro-pointing either. Instead, subjects appeared to have a fuller, more veridical awareness of their actual movement when they made an error compared with either the normal pro-pointing response that the errors resembled, or with the anti-pointing context in which the errors occurred. It seems as if error awareness throws the reality of our actions into a vivid and detailed conscious state, which our movements typically lack.

We suggest this special process of error awareness is at least partly reconstructive. Subjects cannot realise that they have made an anti-point error until after the erroneous deviation towards the target has occurred, since awareness of normal pro-pointing responses lags behind motor performance. Therefore, some process occurring after the initial error must reconstruct the contents of awareness after the error has been registered as such. Computational models of motor control (Wolpert, 1997) offer one possible explanation of error awareness. Such models compare the actual state of an effector to its desired state. When these do not cancel, the motor command is adjusted in proportion to the error signal. We suggest the error signal may also act as a bottom-up gatekeeper to consciousness: when a significant error is detected, sub-cortical circuits, notably in the cerebellum correct the movement. In addition, other circuits bring the error to conscious awareness, to allow a deliberate, voluntary correction, and an explicit phenomenal experience which should favour learning.

'I', Agency and Action

Finally, we would like to suggest that the phenomenology of action has a special contribution to make to consciousness studies. Scientific accounts of consciousness have repeatedly stumbled over the philosophical issue of the relation between primary and secondary consciousness. Put another way, the relation between conscious perceptions and self-consciousness remains unclear. This relation cannot be ignored even in a scientific account of primary consciousness, because all conscious states seem, as part of their content, to imply a self, or 'I' to have the experience. The traditional, reflexive account of self-consciousness explains secondary consciousness by positing a secondary *process* which monitors the primary contents of conscious perception. This view has the attraction of being easy to model computationally: self-consciousness becomes a parallel side-loop receiving copies of the outputs of each primary process (Johnson-Laird, 1983). However, the monitoring account of self-consciousness has the disadvantage of risking an infinite regress of monitoring circuitry. There is also no obvious single neural substrate to perform the monitoring. Moreover, the reflexive account implies the logical possibility that, after some appropriate form of brain damage, a patient should have completely normal experience of all sensory modalities, yet not believe that these experiences belong to them. No such patients have been found. While denial of self-ownership of experience has been reported, such cases tend to be very local delusions, confined to a specific type of primary conscious experience, such as individual body parts (Halligan *et al.*, 1995).

We suggest that the phenomenology of action avoids these problems of reflexivity. In awareness of action, the 'I' is already present, in a much more explicit way than in perceptual states. We have already seen that the awareness of voluntary actions differs dramatically from the awareness of physically-equivalent movements which are imposed involuntarily on the body (Haggard *et al.*, 2002; Tsakiris & Haggard, 2003). In exteroceptive senses, such as vision or audition, the self is present only very implicitly, as the spatial location at the origin of a perceptual scene. Recently, Bermudez has pointed out that a sense of self is much more explicitly attached to primary perceptions in the case of somatosensation, because the sensory signals necessarily refer to one's own body (Bermudez, 1998). We agree with Bermudez that proprioception is a forerunner of self-consciousness. However, we believe that this is even more true of voluntary action than of any perceptual modality, in agreement with Metzinger (2003). We have already seen that intentions, or neural processes preceding actual muscle contraction, contribute substantially to the awareness of action. We contend that awareness of intention has two components. The first is the element of specifying and initiating a determinate movement (Michotte's 'tendance determinante'). The second is the element of source: the experience of action is tagged with 'I' as the source of the action. Introspection tells us that the primary experience of action varies considerably depending on whether 'I' am the source or not: compare the sensation of lifting your arm with the sensation of your arm 'lifting itself' by Kohnstamm's manoeuvre, for example (press the arms outwards against the sides of a doorway for 2 minutes, then stand with your arms held loosely by your sides, and your arms will rise against gravity, 'by themselves').

Future research might study the link between action awareness and the sense of self by attempting to manipulate and combine sources of action in a systematic way. To date, though, quantitative and experimental methods have not found a truly convincing way to operationalise the self in the context of action studies. The link between neuroscience and trusting the subject breaks down at the point of self-consciousness. Perhaps this is because our laboratory experiments tend to focus on the motor performance and motor psychophysics of an individual subject. In these cases, the subject is generally considered more from a third-person perspective than from a first-person perspective, and the content of the experiment does not allow any alternative agent or 'non-self' to highlight the role of the self in voluntary action (but see Tsakiris and Haggard [2003] for an exception). Thus, the psychology of action does 'trust the subject' in the sense of taking their action as a direct and quantifiable consequence of their internal states. However, it remains difficult to relate the actions that we can measure to the subject's conscious experience of agency. Cognitive neuroscience has made enormous progress in recent years in relating specific patterns of brain activity to specific conscious states. The ability to link overt behaviour, brain activity and subjective report offers the best possibility, in the future, to understand the neural and psychological processes of voluntary action.

Until then, the final words on the conscious experience of action may have to be descriptions of subjective experience, developed by the introspectionists of almost a century ago:

> Cette conscience du moi semble même être l'un des premiers critères du phénomène volontaire: c'est la première distinction fait pars les sujets, et elle est très nette des le début.(…) la conscience du moi est intégrée phénoménologiquement dans le cours des faits qui constituent la décision (Michotte & Prum, 1910, p. 133)

> (This self-awareness seems to be one of the primary characteristics of volition: it is the first quality noted by the subject, and is clear from the very start… self-awareness is phenomenologically part of the course of events which make up the process of decision).

To conclude, we have suggested that voluntary actions, though generally neglected in psychology, have a distinctive phenomenology which can be studied both qualitatively and quantitatively. Studies suggest that voluntary actions integrate and bind our conscious experience. This experience may be, in itself, phenomenologically thin, but it contributes to our coherent and unified sense of ourselves as agents, and of a self as the source of our actions.[1]

References

Ach, N. (1905), *Ueber die Willenstätigkeit und das Denken* (Göttingen).
Anscombe, G.E.M. (1963), *Intention* (New York: Cornell University Press).
Bekkering, H. & Wohlschlager, A. (2002), 'Action perception and imitation', in *Common Mechanisms in Perception and Action: Attention and Performance*, ed. W. Prinz & B. Hommel (Oxford: Oxford University Press).
Bermudez, J.L. (1998), *The Paradox of Self-Consciousness* (Cambridge, MA: MIT Press).
Broadbent, D.E. (1982), 'Task combination and selective intake of information', *Acta Psychol*, **50**, pp. 253–90.
Candlish, S. (1981), 'Intention and intentionality', *Philos Quart*, **31**, pp. 170–2.
Castiello, U., Paulignan, Y. & Jeanerod, M. (1991), 'Temporal dissociation of motor responses and subjective awareness: A study in normal subjects', *Brain*, **114**, pp. 2639–55.
Fodor, J.A. (1975), *The Language of Thought* (Cambridge, MA: Harvard University Press).

---

[1] This work was made possible by award of a Leverhulme Research Fellowship. Helen Johnson was supported by a BBSRC Research Committee Studentship. We are grateful to Clare Press for assistance.

Fodor, J.A. (1981), *Representation: Philosophical Essays on the Foundations of Cognitive Science* (Cambridge, MA: The MIT Press).
Haggard, P. (2001), 'The psychology of action', *Brit J Psychol*, **92**, pp. 113–28.
Haggard, P., Clark, S. & Kalogeras, J. (2002), 'Voluntary action and conscious awareness', *Nat Neurosci*, **5**, pp. 382–5.
Haggard, P. & Eimer, M. (1999), 'On the relation between brain potentials and the awareness of voluntary movements', *Exp Brain Res*, **126**, pp. 128–33.
Haggard, P., Newman, C. & Magno, E. (1999), 'On the perceived time of voluntary actions', *Br J Psychol*, **90**, pp. 291–303.
Halligan, P.W., Marshall, J.C. & Wade, D.T. (1995), 'Unilateral somatoparaphrenia after right hemisphere stroke: a case description', *Cortex*, **31**, pp. 173–82.
James, W. (1890), *The Principles of Psychology* (New York: Henry Holt).
Johnson-Laird, P.N. (1983), *Mental Models: Towards a Cognitive Science of Language, Inference, and Consciousness* (Cambridge: Cambridge University Press).
Johnson, H., van Beers, R.J., & Haggard, P. (2002), 'Action and awareness in pointing tasks', *Exp Brain Res*, **146**, pp. 451–9.
Külpe, O. (1893), *Grundriss der Psychologie. Auf experimenter Grunlage dargestellt. [Outlines of Psychology. Based upon the Results of Experimental Investigation]* (Leipzig: Wilhelm Engelmann).
Libet, B., Gleason, C.A., & Wright, E.W. (1983), 'Time of conscious intention to act in relation to onset of cerebral-activity (readiness potential): The unconscious initiation of a freely voluntary act', *Brain*, **106**, pp. 623–42.
Metzinger, T. (2003), *Being No One: The Self-Model Theory of Subjectivity* (Cambridge, MA: MIT Press).
Michotte, A. & Prum, E. (1910), 'Le Choix Volontaire et ses antécédents immédiats', *Arch de Psychol*, **38–39**, pp. 8–205.
Milton, J. (1667), *Paradise Lost, Book IX*.
Mitroff, I. (1998), *Smart Thinking For Crazy Times: The Art of Solving the Right Problem* (San Francisco, CA: Berrett-Koehler).
Reason, J. (1990), *Human Error* (Cambridge: Cambridge University Press).
Searle, J.R. (1983), *Intentionality: An Essay in the Philosophy of Mind* (Cambridge: Cambridge University Press).
Skinner, B.F. (1957), *Verbal Learning* (New York: Appleton-Century-Crofts).
Stout, G.F. (1906), 'The nature of conation and mental activity', *Brit J Psychol*, **2**, pp. 1–15.
Tsakiris, M. & Haggard, P. (2003), 'Awareness of somatic events associated with a voluntary action', *Exp Brain Res*, **139**, pp. 439–46. DOI 10.1007/s00221-003-1386-8.
Wegner, D.M. (2003), 'The mind's best trick: How we experience conscious will', *Trends Cogn Sci*, **7**, pp. 65–9.
Wolpert, D.M. (1997), 'Computational approaches to motor control', *Trends Cogn Sci*, **1**, pp. 209–16.

Shaun Gallagher

Phenomenology and Experimental Design
Toward a Phenomenologically Enlightened Experimental Science

I review three answers to the question: How can phenomenology contribute to the experimental cognitive neurosciences? The first approach, neurophenomenology, employs phenomenological method and training, and uses first-person reports not just as more data for analysis, but to generate descriptive categories that are intersubjectively and scientifically validated, and are then used to interpret results that correlate with objective measurements of behaviour and brain activity. A second approach, indirect phenomenology, is shown to be problematic in a number of ways. Indirect phenomenology is generally put to work after the experiment, in critical or creative interpretations of the scientific evidence. Ultimately, however, proposals for the indirect use of phenomenology lead back to methodological questions about the direct use of phenomenology in experimental design. The third approach, 'front-loaded' phenomenology, suggests that the results of phenomenological investigations can be used in the design of empirical ones. Concepts or clarifications that have been worked out phenomenologically may operate as a partial framework for experimentation.

How can phenomenology contribute to the cognitive sciences? A number of authors have recently raised this question and have proposed diverse answers (see, for example, the essays in Petitot *et al.*, 1999 and Varela and Shear, 1999). I will outline three different responses to this question, with specific reference to the issue of how phenomenology might contribute to experimental design. Although some of these responses will be more positive than others, each one, even the least positive, will be more positive than the fully negative answer proposed by Dennett (2001): 'First-person science of consciousness is a discipline with no methods, no data, no results, no future, no promise. It will remain a fantasy'. The differences between the positions I will outline depend not only on differences in how one understands the role of phenomenology in empirical science, but also on differences in conceptions of phenomenology. Furthermore, in so far as these approaches argue for a naturalized phenomenology, they hold to different conceptions of how naturalization is to be accomplished.

Neurophenomenology

One view, neurophenomenology, as espoused by Francisco Varela (1996), follows Husserl in understanding phenomenology to be a methodologically guided reflective examination of experience. This view maintains that both empirical scientists and experimental subjects ought to receive some level of training in phenomenological method (also see Roy, *et al.*, 1999). Varela proposes that this training would include learning to practice the phenomenological reduction; that is, the setting aside or 'bracketing' of opinions or theories that a subject may have about experience or consciousness. This approach might at first seem methodologically abstract, but Lutz *et al.* (2002) have shown its practicality with some success.

In many empirical testing situations that target specified cognitive tasks, successive brain responses to repeated and identical stimulations, recorded for example by EEG, are highly variable. The source of this variability is presumed to reside mainly in fluctuations due to a variety of cognitive parameters defined by the subject's attentive state, spontaneous thought process, strategy decisions for carrying out the task, etc. For purposes of this paper, let's call these *subjective parameters* and abbreviate this to SPs — they include distractions, cognitive interference, etc. To control for SPs is difficult. As a result, they are usually classified as unintelligible noise (Engel *et al.*, 2001) and ignored or neutralized by a method of averaging results across a series of trials and across subjects. Lutz and his colleagues decided to approach the problem of SPs in a different way. They followed a neurophenomenological approach that combined first-person data and the dynamical analysis of neural processes to study subjects exposed to a 3D perceptual illusion.[1] On the one hand we might think of this study as an attempt to read the subject's experience through a third-person analysis, but Lutz and his colleagues used the first-person data not simply as more data for analysis, but as contributing to an organizing analytic principle.

Specifically, the trials were clustered according to first-person descriptive reports concerning the experience of SPs, and for each cluster separate dynamical analyses of electrical brain activity, recorded by EEG, were conducted. The results were different and significant in comparison to a procedure of averaging across trials.

---

[1] Random dot patterns with binocular disparities (autostereogram) were presented on a computer screen. By visually manipulating these dots, subjects were able to see a 3D illusory geometric shape emerge (depth illusion). They were instructed to press a button with their right hand as soon as the shape had completely emerged. After the button push, the subjects gave a brief verbal report of their experience.

The phenomenological part of the experiment involved the development of descriptions (refined verbal reports) of the SPs through a series of preliminary or practice trials, using a well-known depth perception task. In this preliminary training process subjects became knowledgeable about their own experience, defined their own categories descriptive of the SPs, and could report on the presence or absence or degree of distractions, inattentive moments, cognitive strategies, etc. Based on the subject's own trained reports, descriptive categories were defined *a posteriori* and used to divide the trials into phenomenologically based clusters.[2] Subjects were then able to use these categories during the main trials when the experimenters recorded both the electrical brain activity and the subject's own report of each trial. The reports during the main trials revealed subtle changes in the subject's experience due to the presence of specific SPs, reflecting, for instance, the subject's cognitive strategy, attention level or inner speech. The clustered first-person data were correlated with both behavioural measures (reaction times) and dynamic descriptions of the transient patterns of local and long-distance synchrony occurring between oscillating neural populations, specified as a dynamic neural signature (DNS). Lutz *et al.* cite evidence indicating that such coherent temporal patterns occur during ongoing activity related to top–down factors such as attention, vigilance or expectation. They were able to show that distinct SPs, described in the subjects' trained phenomenological reports, translate into distinct DNSs just prior to presentation of the stimulus. For example, characteristic patterns of phase synchrony recorded in the frontal electrodes prior to the stimulus depended on the degree of preparation as reported by subjects. Lutz *et al.* show that these DNSs then differentially condition the behavioural and neural response to the stimulus.

To be clear, phenomenological training in this experiment did not involve teaching subjects about the philosophical work of Husserl or the phenomenological tradition. Rather it consisted in training subjects to deliver consistent and clear reports of their experience. Trained reflective introspection combined with an attempt to firm up descriptive protocols, based on that reflective stance, may indeed be considered phenomenological training, and as Lutz and his colleagues have shown, it is clearly not impractical. But is it a genuine phenomenological method as Varela describes it — that is, as informed by the phenomenological reduction? How does it differ from other attempts to train introspection?

The goal of the phenomenological reduction is to attain intuitions of the descriptive structural invariants of an experience, not to average them out.

---

[2] For example, with regard to the subject's experienced readiness for the stimulus, the results specified three readiness states: **Steady readiness (SR):** subjects reported that they were 'ready', 'present', 'here', 'well-prepared' when the image appeared on the screen and that they responded 'immediately' and 'decidedly'. **Fragmented readiness (FR):** subjects reported that they had made a voluntary effort to be ready, but were prepared either less 'sharply' (due to a momentary 'tiredness') or less 'focally' (due to small 'distractions', 'inner speech' or 'discursive thoughts'). **Unreadiness (SU):** subjects reported that they were unprepared and that they saw the 3D image only because their eyes were correctly positioned. They were surprised by it and reported that they were 'interrupted' by the image in the middle of an unrelated thought.

Briefly, the reduction involves the bracketing of our ordinary attitudes in order to shift our attention from *what* we experience to *how* we experience it. Varela (1996) identified three steps in phenomenological method, each of which requires training.

(1) suspending beliefs or theories about experience;
(2) gaining intimacy with the domain of investigation;
(3) offering descriptions and using intersubjective validations.

The reduction can be either self-induced by subjects familiar with it or guided by the experimenter through open questions — questions not directed at opinions or theories, but at experience (see Vermersch, 1994 and Petitmengin-Peugot, 1999). Rather than employing pre-defined categories, and asking 'Do you think this experience is like X or Y or Z?' the open question asks simply, 'How would you describe your experience?'

> To train the subjects, open questions were asked to try to redirect their attention towards their own immediate mental processes before the recordings were taken. For example: Experimenter: 'What did you feel before and after the image appeared?' Subject S1: 'I had a growing sense of expectation, but not for a specific object; however when the figure appeared, I had a feeling of confirmation, no surprise at all'; or subject S4: 'it was as if the image appeared in the periphery of my attention, but then my attention was suddenly swallowed up by the shape'. (Lutz *et al.*, 2001).

Open questions posed immediately after the task help the subject to redirect his/her attention towards the implicit strategy or degree of attention he/she implemented during the task. Subjects can be re-exposed to the stimuli until they find 'their own stable experiential invariants' to describe the specific elements of their experiences, in this case the SPs. These invariants then become the defining elements of the phenomenological clusters that are used as analytic tools in the main trials.

The experimental protocol used in Lutz *et al.* (2001) thus employs a practical phenomenological reduction. The subjects are asked to provide a description of their own experience using an open-question format, and thus without the imposition of pre-determined theoretical categories. They are trained to gain intimacy with their own experience in the domain of investigation. The descriptive categories are intersubjectively and scientifically validated both in setting up the phenomenological clusters and in using those clusters to interpret results that correlated with objective measurements of behaviour and brain activity. Of the three approaches reviewed in this paper, this one is the strictest application of phenomenological method in the experimental context.[3]

The Retrospective and Indirect use of Phenomenology

In regard to usual experimental practice, Overgaard (2001) notes: 'Most of the experimental approaches to consciousness simply ignore these issues and will

---

[3] For further theoretical and methodological discussion of this experiment, see Lutz (2002).

either just assume certain experienced qualities in the subject or rely on the more unspecific everyday phenomenology. This is not odd at all when considering the immense work on developing a useful phenomenology that is needed to do this properly' (§ 9). The 'immense work' of such training may be only a *perceived* impracticality, however. Training humans in a reflective procedure (as in Lutz *et al.*, 2002) seems clearly easier than training monkeys in an experimental response mode. Nonetheless, a perceived impracticality is likely to motivate a less formal version of this reflective approach. Braddock (2001) argues for this less formal approach and calls it 'indirect phenomenology'.

Braddock, for example, argues that the practice of phenomenology can be naturalized by allowing results from the cognitive sciences, including, especially, the study of pathological cases, to inform phenomenological analysis, and vice versa. He finds the tradition of Jaspers and Merleau-Ponty to be exemplary in this regard. In principle this seems a good strategy. Someone like Merleau-Ponty (1962) did his phenomenology fully informed by the current scientific research, and used phenomenology, retrospectively, to interpret the results of that science. But this does not address the issue of how to incorporate phenomenology into the experimental situation.

Merleau-Ponty, for example, frequently used phenomenological insights to reinterpret experimental results. In such cases, phenomenology takes on a critical function, offering correctives to various theoretical interpretations of the empirical data. This approach can be theoretically productive in that it develops alternative interpretations. But unless these interpretations are subject to further empirical testing they remain unverified. This simply brings us back to the question of how to incorporate such phenomenological insights into experimental studies.

Another problem with this kind of after-the-fact reinterpretation can be seen in regard to pathological case studies. For example, Merleau-Ponty offers a brilliantly conceived reinterpretation of the case of Schneider, a brain-damaged patient of Goldstein (Merleau-Ponty, 1962). Is the reinterpretation correct? As far as I know, there has never been any attempt to take the phenomenologically inspired reinterpretation back to the laboratory — that is, there was no attempt to translate the phenomenological reinterpretation into any kind of follow-up testing,[4] and as a result, Merleau-Ponty's account of the case remains simply one of several possible theoretical accounts. In a very practical way this suggests the inadequacy of this approach if phenomenologists are not working directly with and along side psychologists and neuroscientists in the experimental context. Once again this brings us back to the question of how specifically to incorporate phenomenology into the experimental context.

Another possible interpretation of indirect phenomenology is that to use phenomenology in experimental testing just means taking introspective reports into consideration. As Braddock himself admits, on this view, what starts out as a conception of formal (e.g., Husserlian) phenomenology, ends up as more or less the kind of informal phenomenology that is currently practiced in the cognitive

---

[4] Indeed, the case of Schneider was an old one when Merleau-Ponty produced his account. Schneider's brain damage was extremely complex, and was studied by Goldstein between 1915 and 1930.

sciences, or something that is akin to what Dennett (1991) calls heterophenomenology. In heterophenomenology, first-person data are averaged out in statistical summaries. That is, first-person data are treated as third-person facts (e.g., behavioural responses) to be analyzed using mathematical instruments and pre-established categories. As a result, no attention is paid to the phenomenal details of the subject's experience. I have suggested (Gallagher, 1997), however, that this procedure is actually naive, and ultimately unscientific, to the extent that in attempting to say something about consciousness or cognitive experience, it fails to acknowledge that its interpretations of phenomenological reports are ultimately, and at least in part,[5] based on either the scientist's own first-person experience, or upon pre-established (and seemingly objective) categories that are ultimately derived from folk psychology or an obscure, anonymous, and certainly non-methodological phenomenology. The intentional stance required for the scientist's interpretation of the subject's report is not itself something that has been scientifically controlled.

This naiveté could be corrected by basing the interpretive categories on a methodologically informed phenomenological analysis. That is, if one could establish the interpretive categories in a phenomenologically controlled way, then the first-person data would not be washed out of the experiment but would be given their proper significance. One way to establish the scientific credentials of the interpretive categories would take us back to a neurophenomenology of the type outlined above.

Neurophenomenology, as we have seen, employs a phenomenological reduction. Although Dennett (2001) introduces his own version of a heterophenomenological reduction, it serves a very different purpose. After cataloguing subjective reports and other first-person data, Dennett suggests that they all be '*bracketed for neutrality*'. In effect he advises the scientist to treat the verbal reports as if they were fiction. One requires, for this process, a suspension of trust in the subject, and a suspicious interpretational practice. Braddock (2001) thinks this is problematic, but in this particular part of the heterophenomenological approach it seems to me to be nothing other than good scientific practice. It involves testing the experiential reports against all the other non-experiential data, and attempting to draw a coherent third-person picture of the subject's behaviour. This is problematic only if one wants to know what the subject's *experience* is like, and what effect that experience might have on the subject's behaviour. A more complete understanding of experience and behaviour, then, does not eliminate the need for a phenomenological analysis to legitimize the initial cataloguing and interpretation of the original subjective reports.

Of course, Dennett's notion of heterophenomenology is motivated by longstanding suspicions about introspection as a psychological method. With the rising importance of brain-imaging techniques, however, there is a renewed

---

[5] In Dennett's most recent version of heterophenomenology he explains that it is not just the verbal reports that constitute the data for heterophenomenological analysis, but behavioural and other objective (physiological) data. So some part of the interpretation of the verbal reports would likely be based on the other objective data.

interest in introspective methodology (see, e.g., Jack and Roepstroff, 2002; Schooler, 2002; and follow-up discussions by Frith, 2002 and Gallagher, 2002). Renewed interest in introspection, however, again directs us to the question of precisely how the use of introspection might be made methodologically secure, that is, how it might be more formally controlled using phenomenological techniques in experimental paradigms.

Each of the various proposals for the *indirect* use of phenomenology leads us back to methodological questions about the *direct* use of phenomenology in experimental design. We have seen that Varela's notion of neurophenomenology, as practiced by Lutz *et al.* (2002), offers one model for such direct use. It is a model that provides a clear proposal for making introspection methodologically secure, and it seems quite possible to use such a procedure in many paradigms that test perceptual consciousness. It is not clear, however, that the specifics of this model can be adapted for use in all experimental investigations of consciousness or cognition. In the following, I consider an alternative model for introducing phenomenology directly into experimental design.

Front-loaded Phenomenology

In this section I want to outline and defend a third view of a phenomenologically enlightened experimental science, or what I'll call simply 'front-loaded phenomenology'. Rather than starting with the empirical results (as one would do in various indirect approaches), or with the training of subjects (as one would do on the neurophenomenological approach discussed above) this third approach would start with the experimental design. The idea is to front load phenomenological insights into the design of experiments, that is, to allow the insights developed in phenomenological analyses (modelled on Husserlian description, or the more empirically oriented phenomenological analyses found, for example, in Merleau-Ponty, or in previously completed neurophenomenological experiments) to inform the way experiments are set up. To front load phenomenology, however, does not mean to simply presuppose phenomenological results obtained by others. Rather it involves testing those results and more generally a dialectical movement between previous insights gained in phenomenology and preliminary trials that will specify or extend these insights for purposes of the particular experiment or empirical investigation. I'll discuss several examples of how this can be done.

Let me begin, however, with two notes. First, and on the one hand, according to this approach, one can incorporate the insights of phenomenology into experimental protocols without training subjects in the method. On the other hand I think that phenomenological insights developed on the basis of such training and in neurophenomenological experimentation can contribute to experimental design by contributing to the phenomenology that can be front loaded into further experiments. That is, the phenomenology that is front loaded might be developed in pure phenomenological analysis (as in Husserl), or in neurophenomenological experiments. Second, it is a natural consequence of front-loading phenomenology that, as in the neurophenomenological model, the phenomenology becomes part of the analytic framework for interpreting the results, and not just part of the data to be analyzed.

The experiments that I will focus on here all involve brain imaging. Furthermore, I have deliberately chosen experiments that do not involve introspective reports in order to eliminate any confusion about where precisely the phenomenological contribution lies on this approach.[6] Specifically, the phenomenological analysis is done prior to the experiments and the results of that analysis are used to work out the experimental design.

For the first two experiments, the phenomenology concerns a distinction between self-agency and self-ownership. In the normal experience of intentional action these two aspects of self-awareness are close to indistinguishable. But consider the phenomenology of involuntary action. If, for example, someone pushes me from behind, I sense that it is my body that is moving — it is *my* movement and I experience ownership for the movement — but I do not experience agency for the movement (I have no sense that I intended or caused the movement). To get the phenomenology right, however, we need to distinguish between the first-order phenomenal level of experience and higher-order cognition. It is possible to make the distinction between 'attributions of subjectivity' (or ownership) and 'attributions of agency' on the level of higher-order, reflective or introspective report (e.g., Graham and Stephens, 1994; Stephens and Graham, 2000). It is also possible to make the distinction at the level of first-order phenomenal consciousness (Gallagher, 2000; 2003a). That is, in the case of involuntary movement, I directly experience the movement as happening to me (sense of ownership), but not as caused by me (no sense of agency). Ownership and agency are seemingly (and in the case of phenomenal experience, 'seemingly' means 'really') built into experience. They are part of a pre-reflective (non-conceptual) self-awareness implicit to the experience of action. Indeed, this is usually the basis for attributions of subjectivity and agency at the higher introspective level.[7]

If neuroscience accepts this phenomenological distinction, then one task is to determine what neurological processes generate these first-order phenomenal experiences. Furthermore, if this distinction is in fact implicit in first-order phenomenal experience rather than the product of second-order introspective attribution, then this suggests that neuroscientists should look for a more basic set of primary processes that are activated in motor control mechanisms rather than in areas that may be responsible for higher-order cognitive processes.

---

[6] There certainly are experiments that rely on front-loaded phenomenology that employ introspective reports. Brøsted (in press), for example, designed an alien-hand experiment to test visual *versus* proprioceptive awareness of the body in bulimic patients, as evidenced by first-person reports. For the experimental design he relies on phenomenological distinctions between body image and body schema (see Gallagher, 1995; Gallagher and Cole, 1995; Paillard, 1997; 1999), and between sense of agency and sense of ownership (Gallagher, 2000a; 2000b).

[7] Graham and Stephens suggest that these distinctions are actually generated at the higher cognitive level on the basis of an intentional or narrative stance that I take toward myself. I've argued that the distinction originally belongs to the first-order level of phenomenal experience (Gallagher, 2003a).

This phenomenologically based supposition has informed the design of several recent experiments. Experimenters have relied on the phenomenological conception of the sense of agency, as distinct from the sense of ownership, as I have defined it, in experiments that attempt to distinguish the neural correlates of the sense of agency for one's own actions (self-agency) in contrast to the sense that the action belongs to someone else (other-agency).

- In the first experiment (Farrer and Frith, 2001), subjects manipulated a joystick to move an image on a computer screen while fMRI brain images were taken. Sometimes the subject caused this movement and sometimes the experimenter caused it. In each case, however, the subject moved the joystick appropriate to the movement on the screen. This allowed for a separation between the sense of agency and the sense of ownership. The effect related to the sense of ownership was present in all conditions and was thus cancelled in the imaging contrasts. The experiments show contrasting activation in the right inferior parietal cortex for perception of action caused by others, and in the anterior insula bilaterally when action is experienced as caused by oneself.[8] The experimenters suggest that the role of the anterior insula in providing a sense of self-agency involves the integration of three kinds of signals generated in self-movement: somatosensory signals (sensory feedback from bodily movement, e.g., proprioception), visual and auditory signals that could generate an ecologically self-specifying sense of movement, and corollary discharge associated with motor commands that control movement. 'A close correspondence between all these signals helps to give us a sense of agency' (p. 602).[9]
- The second study, Chaminade and Decety (2002), involves individual subjects controlling a computer image using a mouse. They are instructed to

[8] Decety et al. (2002), further explore the function of the inferior parietal cortex, and show there is more activation in the left inferior parietal lobule when a subject imitates another person compared to more activation in the right inferior parietal lobule when the other person imitates the subject.

[9] Discussion of these experiments in Farrer and Frith (2001) is not always as clear as it should be in regard to precisely what they were testing. First, in some respects the experimental paradigm, while clearly distinguishing between senses of agency and ownership, fails to distinguish between the first-order phenomenal level and the level of higher-order attribution. I think there are good reasons for interpreting the results in terms of the first-order phenomenal level of experience. It seems reasonable to think that the kinds of information integrated by the anterior insula — proprioceptive feedback, ecological sensory self-specifications involved in movement, and corollary discharge associated with motor commands — constitute implicit, first-order aspects of motor experience rather than the neural correlates of higher-order cognitive attributions or judgments. Second, the fact that a subject moved under both conditions (when she was moving the computer image, and when the computer image was not being moved by her), supposedly, in this context, to distinguish the sense of agency from the sense of ownership, actually confuses the issue with respect to the sense of agency. In one respect it rightly shifts the focus to the issue of the intentional goal of the action — the question is whether I am moving the image on the computer or not. In another respect, however, since at the level of motor behaviour exactly the same movement is made in both cases, it is not clear why that same movement would not generate the same self-specifying information that would tell the subject that she is the agent of that movement. Similar problems are to be found in Farrer et al. (2003). This study, however, nicely shows that conflicts between visual and proprioceptive feedback about one's own movement may cause problems with respect to the sense of agency, and confirms that these problems are correlated with activity in both the posterior insula and the right inferior parietal lobe.

lead or to follow or simply to observe another image on the screen. No reports are required of the subjects; a PET scan images areas of their brains as they perform their movements. The scans show bilateral activity in the inferior parietal cortex in conditions that involved confusion about the origin of an action. When subjects are required to lead (taking the lead in moving an image around on a computer screen), and so when the participant in the scanner sees the other image following his/her actions, more activity is apparent in the right inferior parietal cortex than in the contrasting situation where the subject is asked to follow. In the latter case — involving less of a sense of agency, and more a sense of passive control, or being acted upon — more activation occurs in the left inferior parietal cortex than in the right. The experimenters conclude, 'the lateralization of the inferior parietal lobule activity may be critical for distinguishing consequences of actions generated by the self from those initiated by others, especially when confusion may occur' (p. 1978).

In these experiments subjects can be perfectly naive about the phenomenological details of their own experience. They are not even required to give a report of their experience. Yet it is clear that the phenomenological description of the sense of agency both informs the experimental design (the experiment is set up to find the neural correlates of precisely this experience) and part of the analytic framework for interpreting the results. Moreover, the experiments do not simply presuppose the phenomenological description. Rather, they test and verify that description and extend its application to issues that involve social cognition, at least to the extent that in some of the tasks self-agency is contrasted with the sense that the action is caused by someone else. Since issues pertaining to social cognition and intersubjectivity are also of concern to phenomenology, this opens further opportunities for the interaction between phenomenology and experimental neuroscience.[10] Indeed, in this regard, and as I will suggest in the following example, it is possible to outline a phenomenological proposal for further work in this area.

This final example actually involves a set of important experiments that have already been completed and reported in the literature. They will serve as an example here, but only in the sense that they are missing something that phenomenology could have provided. And what they are missing clearly qualifies their results.

Experiments conducted by Jeannerod, Decety, and their colleagues (Blakemore and Decety, 2001; Decety and Grèzes, 1999; Decety, *et al.*, 2002; Jeannerod, 1997; Ruby and Decety, 2001; and other studies reviewed by Grèzes and Decety, 2001) using various brain-imaging techniques show that there are a

---

[10] Intersubjectivity is an issue in at least two ways in these experiments. First, as mentioned, the experiments target the distinction between self-agency and other-agency, and this can have application to questions of social cognition. Second, the experiments rely upon the interaction between experimenter and subject, and on the instructions or scripts that are presented to the subjects by the experimenters. For an excellent discussion of this issue in regard to top–down versus bottom–up explanations, see Roepstorff and Frith, in press.

number of brain areas (including the SMA, the dorsal premotor cortex, the supramarginal gyrus, and the superior parietal lobe) that are activated in common when a subject

- Engages in intentional action
- Observes others engaging in such action
- Consciously simulates (or imagines) performing such action
- Or prepares to imitate such action

The experimenters who have conducted experiments on these intriguing shared neuronal representations consider the overlapping activation of brain areas to be an important part of the explanation of how we come to understand others. That is, we activate parts of our own motor and cognitive systems in a simulative way, and this neural reverberation gives us insight into what the other person's experience must be like. Those areas that are non-overlapping across these different conditions are also of importance. Jeannerod (2001) has proposed that the non-overlapping areas may account for our ability to distinguish our own activities from those of others, and may contribute to a sense of self-agency (also see Jeannerod *et al.*, 2003; Ruby and Decety, 2001). The experimental paradigm for these experiments is based on an important distinction between first-person perspective and third-person perspective worked out in an influential paper by Barresi and Moore (1996). I've suggested elsewhere, however, that the way in which this distinction is put to use in some experiments suffers from a certain phenomenological impoverishment (Gallagher, 2003b). The two perspectives are defined operationally in the following way in regard to conscious simulation:

First-person perspective: Subjects are asked to imagine themselves performing a given action, for example, reaching to grasp a glass.
Third-person perspective: Subjects are asked to imagine the experimenter performing the same action.

Within the first-person perspective as defined, however, and using a phenomenological technique called 'imaginative variation', I could imagine myself performing an action from within an egocentric spatial framework.

First-person-egocentric perspective: I am located here, and I imagine moving this very hand to grasp the glass in front of me.

Alternatively, however, I could imagine this action using an allocentric spatial framework — taking an external perspective toward myself.

First-person-allocentric perspective: I imagine myself sitting over there, and I can visually imagine how that person, who happens to be me, would reach to pick up a glass that is nearby.

Likewise, for the third-person perspective, it is possible that I could imagine the other person performing the action from an external, allocentric perspective.

Third-person-allocentric perspective: I imagine her over there reaching for the glass.

Or I could imagine taking the other person's place and working out how it must be for her as she reaches for the glass.

Third-person egocentric perspective: I imagine being over there in her place doing the action 'from the inside'.[11]

Table 1: Complex perspectives

Perspectives	First-person	Third-person
Egocentric	'I imagine doing X here'	'I imagine occupying the other's perspective as the other does X'
Allocentric	'I imagine seeing myself doing X over there'	'I imagine seeing the other person doing X'

The issue of perspectives, then, is a complex one, and it leads to the following question: When subjects are asked to simulate (or imagine) performing an action from the first-person perspective (or third-person perspective) do we know whether they are taking an allocentric or egocentric perspective, and is neural activation the same or different across these different perspectives? Employing these phenomenological distinctions and answering this question may help to make the concept of neuronal simulation and the differentiation between self-agency and other-agency more precise.

A further phenomenological, and possibly neurological, complication involves a more precise definition of the mechanisms that allow for the distinction between one's own action, along with the sense of agency for that action, and the actions of someone else. Georgieff and Jeannerod (1998), for example, have proposed a 'Who system' based on the shared (overlapping) and unshared (non-overlapping) neural representations for action. The complication involves what we might call the primary first-person framework that structures all of a subject's experience. That is, in all cases, even in the third-person allocentric framework, I am the one doing the imaginative enactment — third person perspectives are still accomplished within the first-person framework of my own experience. One might say that there is something it is like to be imaginatively enacting an action from a third-person perspective. How is this primary first-person framework accounted for in the 'Who system' so that even when through a process of simulation (personal or subpersonal) I put myself into the place of the other, I never lose track of who is simulating and who is simulated?

These phenomenological distinctions may present difficulties for neuroscience, but it is not beyond the realm of possibilities that they could be front-loaded into the experimental design — specifically by providing instructions to the experimental subjects about how exactly to perform the imaginative enactment. It would also be possible to go further, along the lines of the

---

[11] Farrer and Frith (2001) claim that this is not possible: 'it is not possible to represent the actions of others in the egocentric coordinates used for generating our own actions' (p. 601). It is not clear to me why not.

neurophenomenological procedure of Varela and Lutz, and to train the subjects in a way that would further refine the distinctions. In any case, to the extent that such distinctions are not taken into consideration in experimental design one needs to consider important qualifications on the experimental results and their analysis. The phenomenology of imaginative enactment suggests that the issues concerning overlapping and non-overlapping brain activation is more complex than experimenters may think.

Conclusion

The experiments reviewed in the previous section do not involve any direct introspective reports. I indicated that my choice of experiments was meant to clarify where I think a front-loaded phenomenological approach does its primary work, i.e., in the distinctions and insights that contribute to experimental design and interpretation. The idea is not, I think, that every experiment has to privilege the first-person data internal to the experiment, as long as the significance of first-person experience gets taken into account at some point in the process. In the case of front-loaded phenomenology, the first-person data is taken seriously in the phenomenological analysis that serves to set up the design.[12] In this respect I hope it is obvious that I do not mean to rule out the use of introspective reports in experimentation. To the extent that one may require subjects to report their experiences in the experiment, then the approach of front-loaded phenomenology would be to follow neurophenomenological rather than heterophenomenogical procedures. In all cases, it is good scientific practice to understand and to introduce controls on the various experiential categories that may be involved, both in experiential reports and in the interpretation of those reports. It is not a matter of blindly trusting the subject, or blindly distrusting the subject. Rather, it is a matter of giving both subjects and experimenters methodologically controlled, phenomenologically enlightened ways of understanding the importance of first-person experience and how it can affect the experimental results. Such phenomenologically enlightened approaches are in clear contrast to heterophenomenological procedures that would average out or wash out all first-person data using anonymously formed categories that are considered to be scientific only because they are third-person categories.

Not all scientific concepts are third-person concepts. Consider psychophysical concepts that rely on experience, for example, the felt intensity scales central to Steven's power law (on which the decibel scale for measuring loudness is based). The fact that one cannot eliminate such phenomenal data in well-established areas of scientific psychology has motivated Steven Horst (in press) to remark:

> You simply cannot banish the qualitative aspect of such effects from your description of the psychophysical data: eliminate the qualitative phenomenological property of percept brightness and you have not sanitized the portion of psychophysics concerned with brightness, but eliminated it entirely. No phenomenology, no psychophysics.

---

[12] I thank an anonymous referee for motivating a clarification of this issue.

It is something of a *fantasy*, to use Dennett's term, to suggest that neuroscience or psychology are best done by averaging out, reducing, and re-engineering first-person data so that it looks like third-person data. There is no scientific promise in failing to consider experimental designs that leave the complexity of first-person perspectives out of the equation.

References

Barresi, J., Moore, C. (1996), 'Intentional relations and social understanding', *Behavioral and Brain Sciences*, **19** (1), pp. 107–54.
Blakemore, S.-J., Decety, J. (2001), 'From the perception of action to the understanding of intention', *Nature Reviews: Neuroscience*, **2**, pp. 561–67.
Braddock, G. (2001), 'Beyond reflection in naturalized phenomenology', *Journal of Consciousness Studies*, **8** (11), pp. 3–16.
Brøsted Sørensen, J. (in press), 'The alien-hand experiment', *Phenomenology and the Cognitive Sciences*.
Chaminade, T., Decety, J. (2002), 'Leader or follower? Involvement of the inferior parietal lobule in agency', *Neuroreport*, **13** (1528), pp. 1975–8.
Decety, J., Chaminade T., Grèzes, J., Meltzoff, A.N. (2002), 'A PET exploration of the neural mechanisms involved in reciprocal imitation', *Neuroimage*, **15**, pp. 265–72.
Decety, J., Grèzes, J. (1999), 'Neural mechanisms subserving the perception of human actions', *Trends in Cognitive Sciences*, **3** (5), pp. 172–8.
Dennett, D. (2001), 'The fantasy of first-person science', Nicod Lectures. Private circulation.
Dennett, D. (1991), *Consciousness Explained* (Boston, MA: Little, Brown and Co.).
Engel, A.K, Fries, P., Singer, W. (2001), 'Dynamic predictions: Oscillations and synchrony in top–down processing', *Nature Review of Neuroscience*, **10**, pp. 704–16.
Farrer, C., Franck, N. Georgieff, N., Frith, C.D., Decety, J. and Jeannerod, M. (2003), 'Modulating the experience of agency: a positron emission tomography study', *NeuroImage*, **18**, pp. 324–33.
Farrer, C., Frith, C.D. (2001), 'Experiencing oneself vs. another person as being the cause of an action: the neural correlates of the experience of agency', *Neuroimage*, **15**, pp. 596–603.
Frith, C.D. (2002), 'How can we share experiences?', *Trends in Cognitive Sciences*, **6** (9), p. 374.
Gallagher, S. (in press, 2003a), 'Sense of agency and higher-order cognition: Levels of explanation for schizophrenia', *Cognitive Semiotics*, **2**.
Gallagher, S. (2003b). 'Complexities in the first-person perspective: Comments on Zahavi's *Self-Awareness and Alterity*', *Research in Phenomenology*, **32**, pp. 238–48.
Gallagher, S. (2002), 'Experimenting with introspection', *Trends in Cognitive Sciences*, **6** (9), pp. 374–5.
Gallagher, S. (2000a), 'Philosophical conceptions of the self: implications for cognitive science', *Trends in Cognitive Sciences*, **4** (1), pp. 14–21.
Gallagher, S. (2000b), 'Self-reference and schizophrenia: A cognitive model of immunity to error through misidentification', in *Exploring the Self: Philosophical and Psychopathological Perspectives on Self-experience*, ed. Dan Zahavi (Amsterdam & Philadelphia: John Benjamins), pp. 203–39.
Gallagher, S. (1997), 'Mutual enlightenment: Recent phenomenology in cognitive science', *Journal of Consciousness Studies*, **4** (3), pp. 195–214.
Gallagher, S. (1995), 'Body schema and intentionality', in *The Body and the Self*, ed. J. Bermúdez, N. Eilan, A. Marcel (Cambridge MA: MIT Press).
Gallagher, S., Cole, J. (1995), 'Body schema and body image in a deafferented subject', *Journal of Mind and Behaviour*, **16**, pp. 369–90.
Georgieff, N., Jeannerod, M. (1998), 'Beyond consciousness of external reality: A "Who" system for consciousness of action and self-consciousness', *Consciousness and Cognition*, **7**, pp. 465–77.
Graham, G., Stephens, G.L. (1994), 'Mind and mine', in *Philosophical Psychopathology*, ed. G. Graham, G.L. Stephens (Cambridge, MA: MIT Press).
Grèzes, J., Decety, J. (2001), 'Functional anatomy of execution, mental simulation, and verb generation of actions: A meta-analysis', *Human Brain Mapping*, **12**, pp. 1–19.
Jack, A.I., Roepstroff, A. (2002), 'Introspection and cognitive brain mapping: from stimulus–response to script–report', *Trends in Cognitive Sciences*, **6** (8), pp. 333–9.
Jeannerod, M. (2001), 'Simulation of action as a unifying concept for motor cognition', in *Cognitive Neuroscience: Perspectives on the Problem of Intention and Action*, ed. S.H. Johnson (Cambridge, MA: MIT Press).
Jeannerod, M. (1997), *The Cognitive Neuroscience of Action* (Oxford: Blackwell Publishers).
Jeannerod, M., Farrer, C., Franck, N., Fourneret, P., Posada, A., Daprati, E., Georgieff, N. (in press, 2003), 'Action recognition in normal and schizophrenic subjects', in *The Self and Schizophrenia*, ed. T. Kircher, A. David (Cambridge, Cambridge University Press).
Lutz, A. (2002), 'Toward a neurophenomenology as an account of generative passages: A first empirical case study', *Phenomenology and the Cognitive Sciences*, **1**, pp. 133–67.

Lutz, A., Lachaux, J.-P., Martinerie, J., Varela, F.J. (2002), 'Guiding the study of brain dynamics using first-person data: Synchrony patterns correlate with on-going conscious states during a simple visual task', *Proceedings of the National Academy of Sciences of the USA*, **99**, pp. 1586–91.

Merleau-Ponty, M. (1962), *Phenomenology of Perception*, trans. C. Smith (London: Routledge and Kegan Paul).

Overgaard, M. (2001), 'The role of phenomenological reports in experiments on consciousness', *Psycoloquy*, **12**, p. 29.

Paillard, J. (1999), 'Body schema and body image: A double dissociation in deafferented patients', in *Motor Control, Today and Tomorrow*, ed. G.N. Gantchev, S. Mori, J. Massion (Bulgarian Academy of Sciences. Sofia: Academic Publishing House).

Paillard, J. (1997), 'Divided body schema and body image in peripherally and centrally deafferented patients', in *Brain and Movement*, ed. V.S. Gurfinkel, Yu.S. Levik (Moscow: Institute for Information Transmission Problems RAS).

Petitmengin-Peugeot, C. (1999), 'The intuitive experience', in Varela and Shear (1999).

Petitot, J., Varela, F., Pachoud, B., Roy, J.-M. (ed. 1999), *Naturalizing Phenomenology* (Stanford, CA: Stanford University Press).

Ramsøy, T.Z, Overgaard, M. (2002), 'Breaking the dichotomy: Introspective reports and subliminal perception', poster presented at Association for the Scientific Study of Consciousness. Sixth Conference.

Roepstorff, A., Frith, C. (in press), 'What's at the top in the top–down control of action? Script-sharing and 'top–top' control of action in cognitive experiments', *Psychological Research*.

Roy, J.M., Petitot, J., Pachoud, B., Varela, F. (1999), 'Beyond the gap. An introduction to Naturalizing Phenomenology', in *Naturalizing Phenomenology*, ed. J. Petitot, F. Varela, B. Pachoud, J.M. Roy (Stanford, CA: Stanford University Press).

Ruby, P., Decety, J. (2001), 'Effect of subjective perspective taking during simulation of action: a PET investigation of agency', *Nature Neuroscience*, **4** (5), pp. 546–50.

Schooler, J.W. (2002), 'Re-representing consciousness: dissociations between experience and meta-consciousness', *Trends in Cognitive Sciences*, **6** (8), pp. 339–44.

Stephens, G.L., Graham, G. (2000), *When Self-Consciousness Breaks: Alien Voices and Inserted Thoughts* (Cambridge, MA: MIT Press).

Varela, F. (1996), 'Neurophenomenology : A methodological remedy to the hard problem', *Journal of Consciousness Studies*, **3**, pp. 330–50.

Varela, F., Shear, J. (ed. 1999), *The View from Within* (Exeter: Imprint Academic).

Vermersch, P. (1994), *L'Entretien d'Explicitation* (Paris: ESF).

Bernard J. Baars

How Brain Reveals Mind

*Neural Studies Support the Fundamental
Role of Conscious Experience*

In the last decade, careful studies of the living brain have opened the way for human consciousness to return to the heights it held before the behavioristic coup of 1913. This is illustrated by seven cases: (1) the discovery of widespread brain activation during conscious perception; (2) high levels of regional brain metabolism in the resting state of consciousness, dropping drastically in unconscious states; (3) the brain correlates of inner speech; (4) visual imagery; (5) fringe consciousness; (6) executive functions of the self; and (7) volition. Other papers in this issue expand on many of these points. (Roepstorff; Leopold & Logothetis; Bærentsen; Haggard; Hohwy & Frith).

In the past, evidence based on subjective reports was often neglected (e.g., Ericsson, this issue). It is still true that brain evidence has greater credibility than subjective reports, no matter how reliable. What is new is increasing convergence between subjective experiences and brain observations. For that reason it is no longer rare to see the word 'consciousness' and 'subjectivity' in major science journals. No one so far has discovered a gulf dividing mind and brain. On the contrary, the new evidence supports the central role of consciousness as it was regarded over more than two millenia of written thought.

In a sense this was predictable. Nature is full of unexpected convergences — between fruit fly genes and the human body, between the arc of a tennis ball and the orbit of Mars, and between consciousness and the brain. These convergences show once again the remarkable unity of the observable universe.

I: Introduction

Controversy has marked the scientific study of consciousness for more than a century. Some disputes have been about evidence and many more about philosophy. But scientific controversies are rarely resolved by philosophical reasoning alone.

Many earlier scientific arguments were equally troubling in their time and are now settled. There is general agreement that the earth does go around the sun, that Newton was pretty much right about gravity, and that life is based on carbon chemistry. These questions caused decades of heated debate. Yet somehow consensus emerged. But how? The answer is of course that the path was inductive and empirical, by way of gradually emerging findings and ideas, each one unexpected in its time.

Crucially, the answers that finally emerged were almost always quite different from those that were first envisaged. Early questions are almost always poorly posed, untestable, or based on wrong assumptions.

Consciousness seems to be following a similar path. For 200 years we have collected reliable evidence about sensory consciousness. The decibel scale for sound comes straight from the psychophysics of the 1820s. The color pixels on our screens originated in Newton's experiments in his rooms at Cambridge, showing how a glass prism refracted sunlight into the colors of the rainbow. Both were basic discoveries about conscious perception, because perceived colors and sounds are not in the world. All their perceived properties are in the head. Thus we have a great deal of evidence about consciousness already.

Yet we are still not sure 'what consciousness is' in a deep theoretical sense. This follows the normal history of scientific concepts. Physics in the nineteenth century possessed a wealth of facts about temperature without knowing what temperature really *was*. A workable answer only came with thermodynamic theory, after 1900. In consciousness science our theories also lag behind the evidence, and until a settled theory emerges we will not know what consciousness 'really is'.

The last two decades have shown that conscious brain events are often matched by similar unconscious ones. For example, the meaning of 'this phrase' is unconscious for the reader a fraction of a second before it becomes conscious, even though visual word areas in cortex are already active (Dehaene *et al.*, 2001). After fading from consciousness, the words remain in memory unconsciously for at least ten seconds, as we can tell from sensitive memory tests. The idea that we can often compare conscious and unconscious brain events has profoundly shaped recent science.[1] For the first time we can treat consciousness as an experimental variable, just like any other scientific concept (see Baars, 1997b; Baars, Banks & Newman, 2003). Experimental comparisons make it possible to ask the question, 'What difference does consciousness make in the brain? What does it do for us?'

---

[1] Distinctive and widespread brain activity associated with consciousness has actually been known since 1929, when Hans Berger discovered that the entire brain's electrical activity changes visibly when we wake up. Waking activity reveals electrical voltages that are fast, irregular, and low in amplitude. Deep sleep — the least conscious state of the daily cycle — is marked by voltages that are slow, regular, and much higher in amplitude. Other kinds of unconscious states, such as general anaesthesia, epileptic seizures and coma/vegetative states also show massive slow-wave, high-peak activity. Recent studies indicate that frontoparietal regions are markedly lower in metabolism during unconscious states than in waking control conditions (Baars, Ramsoy & Laureys, under review). Thus we already know that conscious and unconscious states involve distinctive brain-wide patterns of activity.

We can now see much of the living brain at work. Research articles routinely describe brain events reflecting sensory perception, language, mood, mental effort, voluntary control, and much more. In the last decade we have learned an immense amount as a result. (See the classic work of Leopold and Logothetis, Frith and others, this issue).

Obviously the limits of these methods continue to be debated. Yet many scholars now believe that current science is bringing us back to traditional ideas of consciousness, volition, and self that were expelled by behaviorists in the last century. William James would be very much at home with today's ideas. In that sense, studies of the brain reveal the mind with unprecedented clarity and credibility.

II: The Return of Consciousness, Volition and Self

In the year 2000, 1,400 biomedical articles used the word 'consciousness.' In 1950, at the height of behaviourism, there were only five (Baars, 2002a). But in 1950 people were just as conscious as they are today. What has happened is a gradual change in science itself: the slow lifting of the behaviouristic taboo against human subjectivity, which reigned supreme from approximately 1920 to the 1980s.

Following are seven aspects of subjectivity that are now supported by brain observations. Most were beautifully explored in William James' great book, more than 100 years ago (1890).

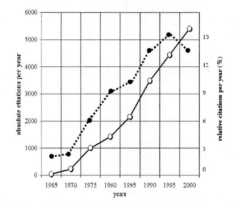

Figure 1. Results of a biomedical database search for the word 'consciousness' and its common synonyms at five-year intervals from 1965 to 2000. The left vertical axis shows absolute numbers of articles (white dots). The right vertical axis shows the number of articles using 'consciousness' (and its synonyms) divided by those using 'behaviour' or 'behavioural', as an index of the size of the overall literature (black dots). The quantitative results are in accord with other sources of information. It appears that the behaviouristic taboo against the word 'consciousness' is no longer dominant.

1. Widespread brain activation during sensory consciousness

How important is consciousness in the brain? Could it be largely unrelated to the working of the nervous system? In the nineteenth century T.H. Huxley suggested that consciousness might be a mere by-product of the brain, with no effect upon it, just as 'the steam-whistle which accompanies the work of a locomotive is without influence upon its machinery' (quoted in James, 1890).

This question now has relevant evidence. For instance, the conventional view of perception is that it involves the brain's detection of sensory scenes, objects and events. And so it does. But there is a great difference between the way the

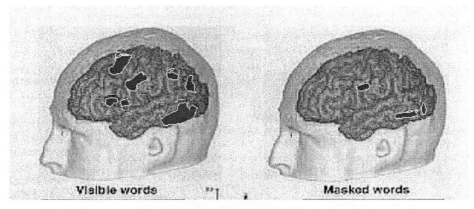

Figure 2. Conscious *versus* masked visual words. Dehaene and co-workers (2001) compared fMRI activation to visible (left) and adjacently-masked (right) words. Dark regions indicate peak averaged hemodynamic activity. Words that were reported as conscious show a 12-fold increase in activation in word recognition regions (fusiform gyrus) relative to masked trials, peaking at 200 ms. Parietal and prefrontal regions show activation only in the conscious and not in the unconscious trials. The scans show group activations in the left hemisphere as seen through a translucent three-dimensional reconstruction of the skull and brain of one of the participants. In these transparent views, the deep activation in fusiform, parietal and mesial frontal cortex appear through the overlying lateral cortices. These results are consistent with current theories of conscious sensory perception (see footnote 2). (With permission from the authors).

brain treats conscious and unconscious sensory perception. When a stimulus is presented unconsciously it activates areas in cortex involved in analysing colours, sounds, faces and the like. But when the identical stimulus is shown consciously, it also recruits regions far beyond the sensory cortex.

Here are three examples out of dozens that have been published (reviewed in Baars, 2002b). Dehaene *et al* (2001) compared conscious words on a screen to the same words when they were masked by a pattern presented immediately before and afterwards. Masked words are unconscious, but they are not physically blocked from entering the eyes. They activate retinal receptors with exactly the same energy pattern as conscious words, and evoke neuronal firing in the visual pathway well into cortex. Thus there is a close similarity between the two stimuli, and it makes sense to compare their brain responses.

Figure 2 shows the results. Unconscious words activated vision and word recognition areas, which analyze such things as stimulus location, shape, colour, and word identity. The identical words, when conscious, triggered 12 times more activity in word recognition regions of the visual cortex. In addition, these words evoked a great deal of additional activity in parietal and frontal cortex. It is as if the stimulus activity in the conscious case is widely distributed from visual regions to other areas in the brain (Baars, 1988; 1997a; 2002b). This is in fact predicted by current theory.[2]

Now consider a very different example, the perception of pain from neurons in the heart region of the body. Rosen *et al* (1996) compared two kinds of neural responses to reduced oxygen supply to the heart. In one group, patients reports intense conscious pain (angina pectoris). The comparison group showed unconscious cardiac anoxia (silent ischemia). Figure 2 shows the results. Again, conscious pain involves very wide activity spreading from the sensory regions to the rest of cortex. By comparison, unconscious 'pain' activity barely reached cortex. As before, similar sensory stimuli have very different effects in the brain, and the difference corresponds to consciousness of the event.

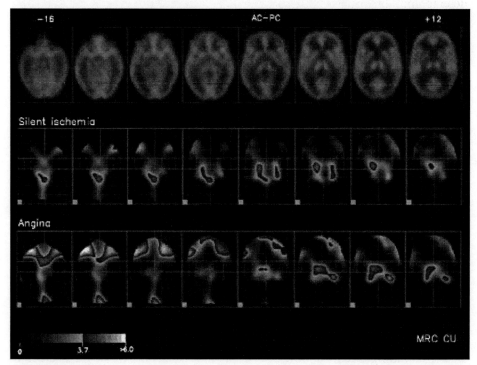

Figure 3. Conscious pain *versus* matched unconscious stimulation. PET was used to measure regional cerebral blood flow changes as an index of neuronal activation during painful and silent myocardial ischemia, induced by intravenous dobutamine. The upper row shows control scans during placebo infusion. The second row shows silent ischemia, with PET activation occurring mostly in bilateral thalami. The bottom row shows significantly higher and more widespread cortical activation during angina (painful ischemia), particularly in the bilateral frontal cortex. Coronal brain slices are viewed from the top, with the eyes looking upward. The leftmost slice is 16 mm below the anterior cingulate (−16). Each succeeding scan to its right is 4 mm higher. (From Rosen *et al.*, 1996, with permission of the authors).

[2] Global workspace theory suggests that consciousness enables brainwide access between otherwise separate functional networks (Baars, 1988, 1997a, 2002b). It is only one of several 'globalist' theories, which interpret consciousness in terms of widespread interactions between many regions in the core brain, regions that are believed to be needed for conscious functions like perceptual continuity, inner speech, imagery, learning and self functions. (Edelman & Tononi, 2000). An increasing number of authors now approach consciousness in terms of a neuronal global workspace capacity. As Daniel C. Dennett recently wrote, 'Theorists are converging from quite different quarters on a version of the global neuronal workspace model of consciousness' (2001, p. 42).

Global Workspace theory has been implemented in a large-scale computer model by Franklin and colleagues. (Baars & Franklin, 2003). It predicts that the specific pattern of distributed activity associated with conscious events should vary from task to task. What is 'global' in such models is the capacity to recruit virtually any set of neuronal resources over a wide range of tasks and conditions.

A recent study by Kreiman et al (2002) adds to these results. Using human patients with electrodes implanted deep in the temporal lobe, to find the source of epileptic seizures, they presented two images to the two eyes. The presentation method, called flash suppression, causes one image to be conscious at a time, though the unconscious image still activates visual cortex. Two-thirds of the deep temporal neurons sampled followed the conscious image; none responded to the unconscious one, even though we know that in visual cortex the unconscious image is represented. Even more important, the temporal regions of the brain do not involve sensory consciousness at all, but rather unconscious aspects of memory and emotion. This shows that conscious events mobilize *unconscious* brain activity outside of the sensory cortex.

Is consciousness a side effect of the brain, like the steam whistle of a locomotive? It seems less and less likely — unless we assume that there is no biological function for the widespread brain activity evoked by the sensory input that is experienced as conscious. The sheer amount of consciousness-related activity suggests a central role for such brain events.

2. High levels of regional brain metabolism in the resting state of consciousness

Our waking hours are taken up with a flow of conscious thoughts, percepts, images, impulses, desires and worries, exertions of will, emotional feelings, inner speech, and intuitions. Yet in spite of sound and reliable self-reported evidence, the stream of thought is rarely studied even today (see Singer, 1993).

That may now begin to change. A revealing brain imaging study was published in 2001 by a French group, Mazoyer and colleagues. These scientists did an unusual thing. Most experiments study the brain during very specific cognitive tasks. Mazoyer and his group turned this around. They asked what would happen if people are simply asked to do nothing, compared to nine standard tasks? Sixty-one subjects in nine PET studies were asked to lie down in the darkened apparatus, to 'keep their eyes closed, relax, refrain from moving, and avoid any structured mental activity such as counting, rehearsing, etc.' What would brain activity and introspective reports be like?

The biggest surprise came when PET scans consistently showed *more* brain activity in the 'rest' condition than in any of nine specific cognitive tasks. Those tasks included imagery, word perception and mental arithmetic. This is quite extraordinary: Whatever subjects were doing while lying in the dark, eyes closed, and trying to relax, requires more metabolic fuel than the standard tasks.

What were they doing? We have two sources of evidence, self-reports and brain activity. Reports from the subjects showed many spontaneous 'autobiographical reminiscences, recent and ancient, consisting of familiar faces, scenes, dialogs, stories, melodies, etc.' Four out of five people reported mental imagery, and three out of four inner speech. About half described mild discomfort from the arm catheter that is required for PET.

The second source of evidence was the PET scans themselves. Mazoyer *et al* found, in technical terms, a network of active brain areas during the rest condition, including 'the bilateral parieto-occipital junction, precuneus, posterior cingulate, and left orbitofrontal cortex'. These areas are involved in:

(1) immediate memory,
(2) control of visual imagery and inner speech
(3) recall of conscious memories
(4) executive functions, and
(5) emotions.

It seems that the spontaneous stream of thought does more important things than mental arithmetic. It is more emotionally driven, more self-involved, more apt to dwell on significant past events, and more likely to attend to plans for the future. It also tends to dwell more on interpersonal conflict (Singer, 1993). By comparison, being told to do mental arithmetic seems only remotely related to the subject's life concerns. Thus the spontaneous stream seems to *do* things for us, even when we aren't sure what it is doing. That may be why the brain is more active during 'rest' than in arbitrary experimental tasks.

Notice how closely the PET scans tracked introspective reports. There is no gulf in this study between objective and subjective evidence. Each is used to interpret the other. Each is revealing in its own way, and together they strengthen our overall understanding.

Does this only prove that subjective reports have been right all along? I believe it does, but it shows more. For one thing, we can now look for unconscious brain activities, the ones people *cannot* report accurately. It is notoriously difficult to introspect on one's own motives, personality, syntax, interpersonal feelings, details of memory processes and automatic habits. These unconscious influences also drive the stream of thought (Luborsky & Crits-Christoph, 1998). Studies like this may also allow us to explore consciousness in people suffering from brain disorders, in babies, and in other animals.

Most important, perhaps, this study shows no gulf between mind and brain. They emerge as two sides of the same mountain.

3. *Inner speech after a century of dispute*

Human beings talk to themselves every moment of the waking day. Most readers of this sentence are doing it now. It becomes a little clearer with difficult-to-say words, like 'infundibulum' or 'methylparabine'. In fact, we talk to ourselves during dreams, and there is even evidence for inner speech during deep sleep, the most unconscious state we normally encounter (Hobson *et al.*, 2000). Overt speech takes up perhaps a tenth of the waking day; but inner speech goes on all the time. According to careful studies, we devote most of our spontaneous inner speech to 'current concerns' (Singer, 1993). Novelists and poets would not find this surprising. But in science, the debate has raged for almost a century whether the introspective evidence was true. Do people really talk to themselves? It is difficult to convey the endless arguments this simple question generated, years after reliable self-reports showed the answer.

Ten years ago the last justification for skepticism quietly crumbled. Paulesu *et al* (1993) found that mental rehearsal of words triggered high activity near the two classic speech regions of the left hemisphere, Broca's and Wernicke's areas, when compared to no mental rehearsal in the same subjects. In 1861 the French neurologist Paul Broca discovered a patient with a stab wound in the brain just in front of the central fissure. It left him unable to speak, but still understanding speech. Some years later, Carl Wernicke, a German doctor, found the complementary area in another patient, a few centimeters behind Broca's area, where local damage made it impossible to understand speech, though the patient could still articulate. These two regions have been implicated thousands of times in speech input and output, but no one could study them in the living brain. Post-mortems were the sole source of brain evidence.

Today we can see with a simple brain scan how inner speech mirrors outer speech, using the same brain regions. The much-debated gulf between mind and brain may be shrinking a little bit.

4. Are visual images like 'faint copies' of sensations?

Mental imagery has a similar history. The reader can experience imagery simply by looking at a light, closing his or her eyes and noticing an afterimage for a few seconds. Or you can imagine the outer door of your house, and ask yourself on which side the doorknob can be 'seen.' It was Aristotle who first suggested that visual images were 'faint copies' of sensations. The American psychologist C.W. Perky showed this elegantly early in the twentieth century, when she demonstrated that people may confuse faint visual pictures with their own mental images (Perky, 1910). Perky used a back-projected screen with a very faint picture of a banana. People were asked to look at the screen and to imagine an object, a fruit perhaps. As Aristotle might have expected, subjects often thought they were *imagining* a banana that was actually faintly in front of them. If we can confuse our own mental images with real visual objects, imagery and vision would seem to be similar. The Perky Effect has now been replicated a number of times (Segal & Fusella, 1970).

This classic experiment might have settled the question once and for all. But behaviorism was on the upswing in the years after 1910 and consciousness was becoming taboo. Many decades later, psychologists began once again to perform experiments on mental imagery, showing that it was quite reliable. But something very odd happened: Imagery was almost never discussed as *conscious*. (Baars, 1988; 1996). Yet we know that people can report mental images, and accurate report is the standard operational index of consciousness.

Brain imaging may be coming to the rescue again. It has shown routinely that the brain areas needed for visual perception are also used in mental images. This is especially true for vivid images — the ones that are most clearly conscious. (Kosslyn, *et al*, 2001; Ishai & Sagi, 1995). Current brain evidence converges so clearly with psychological studies that the most radical skeptics may eventually recognize the facts.

5. Fringe consciousness and the tip-of-the-tongue experience

William James is the fountainhead of consciousness science in the English language, summarizing a century of remarkable studies that were long denounced as unscientific. Many of these have since been verified (James 1890/1983; Baars, Banks & Newman, 2003). It was James who made a surprising observation about 'fringe' or 'vague' experiences in the stream of thought. He strongly argued that fringe experiences are much more common than usually thought, and that they are an essential part of human consciousness. In a famous passage he wrote:

> Suppose we try to recall a forgotten name. The state of our consciousness is peculiar. There is a gap therein; but no mere gap. It is a gap that is intensely active. A sort of a wraith of the name is in it, beckoning us in a given direction, making us at moments tingle with the sense of our closeness, and then letting us sink back without the longed-for term. If wrong names are proposed to us, this singularly definite gap acts immediately so as to negate them. They do not fit into its mold (James, 1890).

The tip-of-the-tongue (TOT) state is an *intention to recall* the missing word, a mental state that lacks perceived or imagined qualities like color, sound, or taste; it has no clear boundaries in space and time, and no contrast between figure and ground. All expectations and intentions seem to be 'non-sensory' in just this sense. To show the power of expectations we need only interrupt some dense flow of predictable experience, for example a printed _____ like this one. Spontaneously we tend to fill in words like 'sentence', anything that fits the meaning and grammar of the sentence. We can see the same effect by interrupting a joke just before the punchline, or a musical tune just before returning to the theme. Clever composers continuously play with our expectations. But the sense of seeking a missing word is not a mental image like your visual image of the cover of this magazine, or the inner sound of *these words*.

James thought that 'fringe' states like this comprise perhaps a third of our mental life; but some of us now believe that they shape *all* of our experiences without exception. The tip-of-the-tongue state provides a nice case to study, because it draws out a colorless expectation over many seconds.

It is easy enough to induce tip-of-the-tongue states. All we need to do is to ask someone for a difficult word – what do you call the flying dinosaurs? What is the capital of Estonia? Hundreds of studies have been published on fringe experiences. But there is a curious omission in this scientific literature. There is almost no exploration of the *subjective experience* of the fringe. Even though everyone quotes James' phenomenological description, it has gone virtually untouched.

Consider a very simple hypothesis. Fringe experiences like the tip-of-the-tongue state lack perceptual qualities, like colour, location or timbre. One reasonable guess, therefore, is that it is a non-sensory state. The human cerebral cortex — the source of conscious contents — can be divided into two great portions. The posterior part is largely sensory. But the frontal third has hardly any sensory areas at all — it deals with motor control, working memory, executive functions, impulse control, aspects of emotion, and the like. Could it be that fringe experiences simply mobilize the front of the cerebrum, and that object-like conscious experiences involve sensory regions in the posterior half?

We now have the first brain imaging study of the tip-of-the-tongue experience (Maril *et al.*, 2001). During TOT states compared to successful word retrieval, Maril and coauthors found, the lateral prefrontal cortex and anterior cingulate are highly activated. This is exactly what we would expect, given the nonsensory 'fringe' nature of that state.

Notice how this study gives us a way to think about the phenomenology of the fringe. No longer do we have to avoid this pervasive aspect of reality. We can ask questions about it, test reasonable hypotheses, and hope to learn more. The brain seems to reveal the mind in remarkable detail.

6. Executive functions of the self — another taboo bites the dust

For half a century sophisticated scientists and philosophers rejected the idea of 'self' as illogical and unscientific. Yet these people were as self-serving as other humans. When criticized they presumably experienced anger and defended themselves. When threatened by professional politics they probably sometimes used the familiar ego defenses we see in the news every day — rationalization, intellectualization of emotion, denial, displacement of blame, the lot. When asked to pass the sugar, they exercised executive control over their skeletal muscles, using prefrontal cortex. The same region of cortex is used to suppress impulsive actions, emotions and appetites, the ones we experience as 'not the kind of thing "*I*" would do'. All these standard aspects of self regulate all of our lives. Yet when asked about 'self,' philosophers and scientists would denounce it as absurd. Some still do.

Gilbert Ryle is generally cited for his critique of self as a 'Ghost in the machine' (1949). Ryle attacked what he claimed was a widespread misconception of self as a homunculus, a little man in the brain. He pointed out that such an observing homunculus could not *explain* self, since it only required another little observer inside itself to make sense of its own experiences, and so on ad infinitum. This critique is still popular.

Most scientific conceptions of self, however, do not involve a homunculus at all (e.g., Baars, 1988; Hilgard, 1977; Dennett, 2001). Ideas of self in neurology, cognitive neuroscience, social psychology, psychodynamics, and personality theory are not homunculi. But Ryle became famous at a time when the whole vocabulary of human psychology was being erased. Self was one more idea for the rubbish heap.

Daniel C. Dennett, who studied with Ryle, has answered the homunculus critique by pointing out that 'an analysis of the Subject (is) a necessary component of any serious theory of consciousness.' (2001, p. 221) As long as the self is accounted for in terms that are not just other selves, Dennett sees no problem. In that sense he is in agreement with scientists who have studied ego functions for the last century.

Brain imaging has also come to the rescue. Neurologists have long known that damage to the front edge of the cortex can result in profound alterations of personality. Most famous is the classic case of Phineas Gage, the nineteenth-century American railroad foreman who was injured by a tamping spike driven explosively through the orbit of his left eye and frontal brain. As Damasio et al. (1994) point out, Gage's doctors were surprised by his robust physical health. But his personality changed. From a model worker Gage turned into an angry drifter, unable to plan for his future or to control impulsive feelings. Such personality changes are common in frontal lobe damage.

Today we can see brain images of prefrontal damage. Even better, we can watch frontal executive regions doing their job in normal, healthy people. In the last few years the literature on the topic has grown so quickly that a search for 'executive prefrontal' shows 289 empirical articles. They include such 'self'-related topics as voluntary decision-making, emotion and motivation, criminal behavior, schizophrenia, and much more.

Self functions engage more than just frontal cortex. But for humans, that part of the brain is involved in personality traits such as persistence and self-control, postponing immediate gratification, moral commitment, and even conscious emotional feelings.

Today the question of the self is back in full force.

7. *The rediscovered problem of volition — not free will*

In the nineteenth century psychology focused on three topics: cognition, volition, and emotion (Hilgard, 1977). To volition William James devoted a remarkable, fact-filled chapter. James discussed such things as people with 'inhibited will' (now called dysexecutive syndrome), 'explosive will' (now called impulse-control disorder), and the like. He also provided a subtle and beautiful theory of voluntary control, called the ideomotor theory, which is still highly plausible (James, 1890/1983; Baars, 1988; 1993; Franklin, 2002). James' ideas were of course erased during the decades of behavioristic dominance.

In the late twentieth century the word 'cognition' and even 'emotion' returned to science, but 'volition' is still a bit lost, in spite of massive evidence (Baars, 1988; 1992; 1993). Indeed, many scientists still claim — voluntarily, of course — not to believe in volition at all (e.g. Wegner, 2001).

The free will debate continues to muddy these waters. Voluntary actions are usually accompanied by a sense of free will. People value that sense of freedom, and may even kill or die for it. It is of fundamental importance. But the *metaphysical* question of free will is a very different matter. It is hard even to state coherently. To conflate volition with free will is to make it impossible to study. For scientists, a useful working assumption is that the feeling of freedom is a deeply held human experience, but that our actions are determined by some causal network like any other. That allows us to explore it empirically.

In sum, the conscious *sense* of free will is real and important, but metaphysical free will is a recipe for endless, useless debate.

III: Brain Studies of Volition

Volition can be studied as an empirical question, as fundamental as consciousness itself. The evidence for both is immense. Major regions of the nervous system are dedicated to voluntary control, especially in prefrontal cortex. Other divisions of the nervous system are independent of voluntary control, such as the autonomic nervous system (so called because it is normally autonomous — free from voluntary control). Billions of neurons control nonvoluntary aspects of action in the cerebellum and basal ganglia. Clinical neurologists, who deal with brain damaged patients every day, have never erased the term 'voluntary' from their vocabulary. The evidence is just too obvious.

Recent brain studies show robust effects of volition. For example, one can show a striking difference in the brain between voluntary swallowing and spontaneous swallowing in humans. Like other vital functions (breathing, for example), swallowing has long been believed to have a dual brain control system. We can do it either voluntarily or spontaneously. If we are trying to swallow a big pill, voluntary effort may be needed. That difference is now seen in brain scans. (Kern, Jaradeh *et al.*, 2001).

There are many other 'double dissociations' between voluntary and involuntary brain events, showing that they do not have the same neuronal basis. For instance, Iwase *et al.* (2002) showed that voluntary smiles show quite different cortical activity compared to spontaneous (unintended) smiles. Yet the behaviour of smiling is similar in both cases.

The famous knee jerk reflex is experienced as involuntary. One can try to imitate it voluntarily with the same spring-loaded dynamic acceleration. But it feels different: People may be surprised by their own reflexes; we are rarely surprised by our voluntary decisions and actions.

These facts mean that volition can be treated as a variable, just like consciousness. (Baars, 1988; 1992; 1997a; 2002; Baars, Banks & Newman, 2003). That is, we can easily compare similar voluntary and involuntary actions. Scientifically this is the only way to study any topic.

Volition is not one problem, but a set of empirical questions. Recent brain studies show progress in understanding the voluntary sense of effort, for example. This is now believed to be the common factor in fluid intelligence tests, the ability to deal flexibly with difficult problems. In a remarkable meta-analysis of many different brain studies, Duncan & Owen (2000) have shown that the same part of the brain is activated with mental effort, regardless of the specific task involved. Since mental effort is a fringe experience (without sensory consciousness), it is not surprising that it activates the dorsolateral prefrontal cortex.

It is impossible to review the relevant studies here. But the sheer number of brain experiments on volition tells the story. Like consciousness and self, the question of voluntary control is back on the frontier.

IV: The Myth of Introspectionism

For many years, the study of conscious experience was derided as 'introspectionism'. Psychology students were taught that introspection had been tried in the nineteenth century and had failed. In fact, that critique was part of the behaviouristic myth of origins propagated in the early twentieth century (Blumenthal, 1979; Danziger, 1979; Hilgard, 1977; Baars, 1986; 2003; in press). Major figures in earlier psychology, like Wilhelm Wundt, vehemently criticized introspective methods as empirically unreliable. Yet Wundt was labeled the ultimate introspectionist, a cruel irony. The supposed failure of consciousness science was meant to justify behaviouristic prohibitions against subjective experience, even when tested and validated objectively. The result was poor science, wrong history, and dehumanized education.

In fact, the nineteenth century discovered the majority of mental phenomena we know today. Human consciousness has been studied scientifically with great success at least since the 1820s. Yet this misleading history continues to be taught.

Reports of reliable conscious experiences are much like other empirical indices in the history of science. They have pros and cons, and are continuously being improved methodologically. New brain-based measures also are emerging. Such a standard scientific effort needs no defense, as long as experiential reports are carefully collected under optimal conditions, and verified in a variety of ways. They are extraordinarily useful. The evidence they provide is often accurate, cumulative, empirical, theoretically elegant, and humanly significant.

V: Where Are We Going? Toward a Neo-Jamesian Science

Science and its technological offspring are unpredictable. But if the evidence arrayed in this paper is roughly true, we may be on the verge of a historic understanding of our own minds, both the conscious and unconscious sides.

Human beings, it is said, cannot stand too much reality. We are our own worst observers. Can we tolerate an accurate understanding of our own minds? It is not at all obvious, and we can imagine bad as well as good ripple effects. Hot debates will no doubt arise about privacy and personal control in a world where our consciousness is no longer walled off from public knowledge. Nightmare scenarios can be found in any science fiction library.

I prefer to hope for a more positive future, a neo-Jamesian science in which scientific knowledge reveals our own lives and makes us more deeply humane. Predictions of disaster do not have a strong track record. The end of privacy was forecast when computers started to become powerful, but in fact, so far the personal computer has vastly expanded the access of average people. Genetic breakthroughs were greeted with dire predictions that have yet to come true. Pandora's curse may be seen in the uncontrolled spread of nuclear science; but on average, the ripple effects of fundamental new findings seems positive.

The truth is that any group merciless enough to use torture already has access to the private thoughts of its victims. Perhaps half the governments in the world resort to such methods. The twentieth century was simply filled with horrific examples of whole populations for whom thinking was not private and not free. Arguably the behaviouristic conception of human beings bereft of consciousness and volition contributed to a mindset in which people became nothing but puppets, to be manipulated at will.

In a neo-Jamesian universe ethical questions become obvious. If we can literally see the pain we inflict upon animals, on babies and perhaps foetuses and each other, the dilemmas can no longer be rationalized and evaded. That may not make life easier, but it makes it more honest. Ultimately, consciousness is a piece of reality, and by and large, we are better off trying to understand it than to evade it.[3]

References

Baars, B.J. (1986), *The Cognitive Revolution In Psychology* (New York: Guilford Press).
Baars, B.J. (1988), *A Cognitive Theory of Consciousness* (Cambridge: Cambridge University Press).
Baars, B.J. (1992), *Experimental Slips and Human Error: Exploring the Architecture of Volition* (New York: Plenum Press).
Baars, B.J. (1993), 'Why volition is a foundation issue for psychology', *Consciousness & Cognition*, **2** (4), pp. 281–309.
Baars, B.J. (1996), 'When are images conscious? The curious disconnection between imagery and consciousness in the scientific literature', *Conscious Cogn.*, **5** (3), pp. 261–4.
Baars, B.J. (1997a), *In the Theater of Consciousness: The Workspace of the Mind* (New York: Oxford University Press).
Baars, B.J. (1997b), 'Treating consciousness as an empirical variable: The contrastive analysis approach', in *The Nature of Consciousness: Philosophical Controversies*, ed. N. Block, O. Flanagan, G. Guzeldere (Cambridge, MA: Bradford/ MIT Press).
Baars, B.J. (2002a), 'Recovering consciousness: A timeline', *Science and Consciousness Review* (www.sci-con.org).
Baars, B.J. (2002b), 'The conscious access hypothesis: Origins and recent evidence', *Trends in Cognitive Sciences*, **6** (1), pp. 47–52. http://www.nsi.edu/users/baars/BaarsTICS2002.pdf
Baars, B.J. (2003), 'The double life of B.F. Skinner', *Journal of Consciousness Studies*, **10** (1), pp. 5–25.
Baars, B.J. (in press), 'I.P. Pavlov and the freedom reflex', *Journal of Consciousness Studies*.
Baars, B.J. & Franklin, S. (2003), 'How consciousness interacts with working memory', *Trends in Cognitive Sciences*, **7** (4), pp. 166–71. http://www.nsi.edu/users/baars/
Baars, B.J., Ramsoy, T.Z. & Laureys, S. (under review), 'Globalist theories of (un)consciousness: Recent evidence'.
Baars, B.J., Banks, W.P. & Newman, J. (2003), *Essential Sources In the Scientific Study of Consciousness* (Cambridge, MA: MIT Press; http://mitpress.mit.edu/0262523027).
Blumenthal, A.L. (1979), 'Wilhelm Wundt, the founding father we never knew', *Contemporary Psychology*, **24**, pp. 547–50.
Damasio, H., Grabowski, T., Frank, R., Galaburda, A.M., Damasio, A.R. (1994), 'The return of Phineas Gage: Clues about the brain from the skull of a famous patient', *Science*, **264** (5162), pp. 1102–5.
Danziger, K. (1979), 'The positivist repudiation of Wundt', *Journal of the History of the Behavioral Sciences*, **15**, pp. 205–26.

---

[3] The author gratefully acknowledges support from The Neurosciences Institute and The Neurosciences Research Foundation. 10640 John Jay Hopkins Drive, San Diego, CA 94549. www.nsi.edu

Dehaene, S., Naccache, L., Cohen, L., Bihan, D.L., Mangin, J.F., Poline, J.B., Riviere, D. (2001), 'Cerebral mechanisms of word masking and unconscious repetition priming', *Nat Neurosci,* **4** (7), pp. 752–8.
Dennett, D.C. (2001), 'Are we explaining consciousness yet?', *Cognition*, **79**, pp. 221–37.
Duncan, J. & Owen, A.M. (2000), 'Common regions of the human frontal lobe recruited by diverse cognitive demands', *Trends Neurosci.*, **23** (10), pp. 475–83.
Edelman, G.M. and Tononi G. (2000), *A Universe of Consciousness: How Matter Becomes Imagination* (New York: Basic Books).
Hilgard, E.R. (1997), *Divided Consciousness: Multiple Controls In Human Thought and Action* (New York: Wiley).
Hobson, J.A., Pace-Schott, E.F., Stickgold, R.. (2000), 'Dreaming and the brain: Toward a cognitive neuroscience of conscious states', *Behav Brain Sci.*, **23** (6), pp. 793–842.
Ishai, A. & Sagi, D. (1995), 'Common mechanisms of visual imagery and perception', *Science*, **268**, pp. 1772–4.
Iwase, M., Ouchi, Y., Okada, H., Yokoyama, C., Nobezawa, S., Yoshikawa, E., Tsukada, H., Takeda, M., Yamashita, K., Takeda, M., Yamaguti, K., Kuratsune, H., Shimizu, A., Watanabe, Y. (2002), 'Neural substrates of human facial expression of pleasant emotion induced by comic films: A PET Study', *Neuroimage*, **17** (2), pp. 758–68.
James, W. (1890), *The Principles of Psychology* (New York: Holt).
Kern, M.K., Jaradeh, S., Arndorfer, R.C., Shaker, R. (2001), 'Cerebral cortical representation of reflexive and volitional swallowing in humans', *Am J Physiol Gastrointest Liver Physiol*, **280** (3), pp. G354–60.
Kosslyn, S.M., Ganis, G. & Thompson, W,L. (2001), 'Neural foundations of imagery', *Nat Rev Neurosci.* **2** (9), pp. 635–42.
Kreiman G, Fried I, Koch C. (2002), 'Single-neuron correlates of subjective vision in the human medial temporal lobe', *Proc Natl Acad Sci U S A.* Jun 11, **99** (12), pp. 8378–83.
Logothetis, N.K. and D.A. Leopold (1998), 'Single-neuron activity and visual perception', in *Toward a Science of Consciousness II*, ed. S. Hameroff, A. Kasznjak, A. Scott (Cambridge, MA: MIT Press).
Luborsky, L. and Crits-Christoph, P. (1998), *Understanding Transference: The Core Conflictual Relationship Theme Method*, 2nd Edition: Contents (Washington, DC: APA Books).
Maril, A., Wagner, A.D. & Schacter, D.L. (2001), 'On the tip of the tongue: An event-related fMRI study of retrieval failure and cognitive conflict', *Neuron*, **31**, pp. 653–60.
Mazoyer, B., Zago, L., Mellet, E., Bricogne, S., Etard, O., Houde, O., Crivello, F., Joliot, M., Petit, L., Tzourio-Mazoyer, N. (2001), 'Cortical networks for working memory and executive functions sustain the conscious resting state in man', *Brain Research Bulletin*, **54** (3), pp. 287–98.
Milner, A.D. & Goodale, M.A. (1995), *The Visual Brain in Action* (Oxford: Oxford University Press).
Paulesu, E., Frith, C.D. & Frackowiak, R.S. (1993) The neural correlates of the verbal component of working memory. *Nature*, **362** (6418), pp. 342–5.
Perky, C.W. (1910), 'An experimental study of imagination', *American Journal of Psychology*, **21**, pp. 422–52.
Rees, G., Kreiman, G., Koch, C. (2002), 'Neural correlates of consciousness in humans', *Nat Rev Neurosci.*, **3** (4), pp. 261–70.
Rosen, S.D., Paulesu, E., Nihoyannopoulos, P., Tousoulis, D., Frackowiak, R.S., Frith, C.D., Jones, T., Camici, P.G. (1996), 'Silent ischemia as a central problem: Regional brain activation compared in silent and painful myocardial ischemia', *Ann Intern Med*, **124** (11), pp. 939–49.
Ryle, G. (1949), *The Concept of Mind* (London: Hutchinson).
Segal, S.J. & Fusella, V. (1970), 'Influence of imaged pictures and sounds on detection of visual and auditory signals', *Journal of Experimental Psychology*, **83** (3), pp. 458–64.
Singer, J.L. (1993), 'Experimental studies of ongoing conscious experience', in *Experimental and Theoretical Studies of Consciousness*. Ciba Foundation Symposium 174, ed. G.R. Bock & J. Marsh (London: Wiley Interscience).
Wegner, D. (2002), *The Illusion of Conscious Will* (Cambridge, MA: MIT Press).

David A. Leopold, Alexander Maier and Nikos K. Logothetis

Measuring Subjective Visual Perception in the Nonhuman Primate

Understanding how activity in the brain leads to a subjective percept is of great interest to philosophers and neuroscientists alike. In the last years, neurophysiological experiments have approached this problem directly by measuring neural signals in animals as they experience well-defined visual percepts. Stimuli in these studies are often inherently ambiguous, and thus rely upon the subjective report, generally from trained monkeys, to provide a measure of perception. By correlating activity levels in the brain to this report, one can speculate on the role of individual neurons and groups of neurons in the formation and maintenance of a particular percept. However, in order to draw valid conclusions from such experiments, it is critical that the responses accurately and reliably reflect what is perceived. For this reason, a number of behavioural paradigms have been developed to control and evaluate the truthfulness of responses from behaving animals. Here we describe several approaches to optimizing the reliability of a monkey's perceptual report, and argue that their combination provides an invaluable approach in the study of subjective visual perception.

Everybody's got something to hide except me and my monkey

John Lennon and Paul McCartney (1968)

Introduction

How the activity of neurons in the brain can give rise to a percept that is 'subjective' has been, and continues to be, one of the great mysteries of science. The faculty of vision has been central in approaching this question both experimentally and theoretically (Crick & Koch, 2003). Modern experimental approaches to understanding neural mechanisms of visual perception often rely on comparing two inherently different quantities. The first is one of many sorts of neurophysiological signals, typically collected from an electrode embedded among neurons in the brain itself. The second is the visual percept generated by the same brain, and is often provided by the report of a trained macaque monkey — a species whose visual abilities are in most respects identical to those of humans. Of these parallel measurements, the latter is often considerably more difficult to obtain, and its accuracy generally limits the validity of any conclusions drawn. Here we review a number of methods that have been applied to overcome the manifest problem of limited access to personal experiences, as well as the further complication of communicating with a different species.

In approaching this issue, it is important to realize that the brain's own action forces upon us a distinction between sensation and subjective perception. Visual 'sensation', as we use the term here, is the reception of a pattern of light on the retina and the automatic, indifferent cascade of neural responses that follows. 'Perception', on the other hand, we reserve for describing what we actually see — or at least what we think we see. Unlike sensation, perception it is an active and interpretive process, drawing upon the benign sensory traces and culminating in our ultimate subjective impression of the visual world. Of course, this dichotomy, like most, is far too simple to be useful. Our sensory machinery is most certainly not impartial to our perception, since it is evolutionarily tuned to facilitate particular percepts and exclude others. And our perception does not simply draw upon our sensory apparatus, but can also shape it according to expectations, the relevance of particular objects, or recent experience. However,

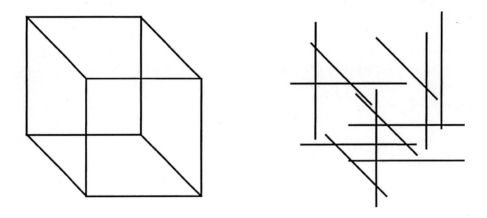

Figure 1

The specific placement of 12 line segments determines how each of them is perceived. (a) By configuring the lines into the well-known Necker cube, each line can take on exactly two perceived configurations. This is a notable reduction from a single line, appearing alone on the sheet, which can, in theory, give the appearance of any of an infinite number of angle/length combinations. (b) Arranging the same segments randomly gives an impression that is much more flat, as if the segments are lying on a flat piece of paper. In some respects, the percept in (b) thus is more veridical than that in (a).

despite these concessions, it is difficult to discard this dichotomy completely for one simple reason: subjective perception can, on occasion, be completely uncoupled from sensory events. One need only consider visual perception during dreaming.

This uncoupling serves as an entry point for research into perception, and visual scientists have therefore often exploited ambiguous patterns that offer more than one valid perceptual interpretation. The ambiguous, and often wavering, impression of these patterns, perhaps more than any other demonstration, illustrates the active and interpretive nature of perception. The famous Necker cube (Figure 1a), for example, offers exactly two valid configurations, with its front face aimed either upward or downward. But why is the Necker cube ambiguous? Careful consideration reveals that its specialized structure does not create ambiguity, but instead constrains it. Each of its 12 line segments, if considered alone, can be conceived in multiple geometries, since an infinity of angle/length combinations in three dimensions give rise to the same two-dimensional projection onto the receptors tiling the retina. Only when these line segments are positioned such that they begin to interact is the interpretation constrained further by the brain. In the case of the Necker cube, exactly two valid interpretations remain. Interestingly, neither of these interpretations can be described as 'veridical', which would hold the line segments to lie on a flat surface. Such a percept can be easily achieved when the same components are arranged in a different manner (Figure 1b). Why then does the brain reject the veridical (flat) interpretation of the Necker cube in favour of a virtual three-dimensional figure that is itself ambiguous? And where in the brain is the ultimate geometry decided? Intriguing questions such as these have been asked with increasing frequency by neurophysiologists, in the hopes that they will provide insight into fundamental mechanisms of visual perception. It is, for example, of great interest to learn how a 'sensory' neuron in the visual cortex (i.e., one whose activity is reliably modulated by the appearance and removal of a visual pattern) would respond under conditions in which the same pattern is *perceived* in two different ways. Does the ultimate perceptual interpretation of a stimulus depend upon (or perhaps contribute to) the activity of such visually responsive neurons? And if so, what types of neurons might show such perception-related activity? To test these questions experimentally in monkeys it is necessary to gain accurate knowledge of their perceptual state.

A number of important differences exist in the collection of psychophysical data from humans and monkeys. In addition to the well-established performance fluctuations and criterion changes that can cause difficulties in measuring human perception, monkeys pose an additional difficulty: they have no *a priori* desire to be 'truthful' in their responses. While the default assumption for humans is that they are responding as accurately as they can, that for monkeys is that they are responding inaccurately — and the investigator must continually prove otherwise. In performing tasks in the laboratory, monkeys are motivated by immediate reward, say a drink of apple juice or a marshmallow, and their main objective is to optimize the frequency of this reward. While this might seem benign, it becomes a problem as soon as the monkey must report its internal, subjective percept with stimuli that are truly ambiguous, since it is no longer possible to differentially reward trials based on their 'truthfulness'. And a monkey who understands that there are no negative consequences for incorrect answers eventually loses motivation to respond accurately.

The present article thus focuses on various tricks and methods currently used to define and evaluate the perception of ambiguous patterns. The ultimate goal of many of these tricks is to shape and maintain an accurate subjective report on the part of the monkey, since this provides the ultimate access to the monkey's perceptual state. We consider two types of multistable stimuli that have been studied in our laboratory. The first is the phenomenon of binocular rivalry (BR), which can be elicited by simply presenting dissimilar stimuli to corresponding regions of the two eyes (Dutour, 1760). Normally the eyes see roughly (but not exactly)

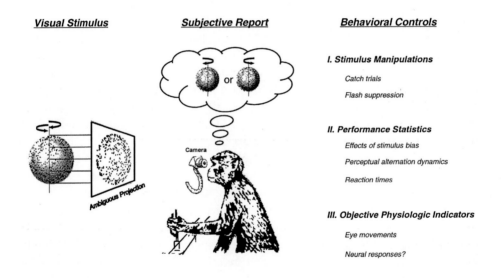

Figure 2

Monkeys are often trained to report their perception of an ambiguous pattern by means of pulling a lever, as shown here. In this example, the pattern is the two-dimensional (orthographic) projection of a transparent, textured sphere. When in motion, this stimulus gives the striking percept of a three-dimensional ball. However, the flat projection does not provide information as to whether the leftward or rightward moving dots represent the front face (e.g., if, as considered from above, the sphere is spinning in a clockwise or counter-clockwise manner). Thus, the brain imposes a particular depth organization onto this ambiguous pattern, ultimately generating the percept of a sphere spinning in one of the two possible directions. As in binocular rivalry and other multistable stimuli, continuous viewing of this pattern gives rise to a dynamic percept, where perceptual dominance (here: clockwise vs. counter-clockwise motion) alternates every few seconds between viable visual interpretations. Given the reliance on the subjective report, a number of approaches are used in order to evaluate and control the monkey's accuracy. These include manipulations biasing interpretation of the stimulus, statistical evaluation of the animal's reported percept, and using objective physiological measures such as eye movements. These approaches are discussed in detail in the text.

the same image of the world. However, when two very different monocular patterns are shown, the brain selects only one to reach conscious perception at any point in time. The second stimulus, shown in the left portion of Figure 2, is comprised of a circular region filled with random dots that, when set in coordinated motion, gives rise to the perception of an illusory three-dimensional rotating sphere (RS). This so-called structure-from-motion stimulus is a striking example of how the brain can use a single cue (relative motion in this case) to construct the impression of depth and three-dimensional shape (Wallach & O'Connell, 1953). The RS stimulus is inherently ambiguous, with either the leftward or rightward dots being perceived as moving on the front surface. Both the BR and RS stimulus, like virtually all ambiguous patterns, destabilize perception. The subjective impression alternates unpredictably between the two solutions as long as the stimulus is present.

We consider here three general approaches for shaping and evaluating the perception of these and related stimuli, outlined on the right in Figure 2. First, it is often possible to physically disambiguate such patterns to provide a coarse measure of the monkey's honesty or reliability. Next, a statistical analysis of the animal's perceptual report can often provide further evidence regarding the accuracy of the monkey's responses. Third, there exists a small number of objective physiological indicators that can provide a trial-by-trial evaluation of the monkey's subjective performance. Below, we give examples of each of these methods, and describe how we have combined these methods to evaluate the monkey's response accuracy in the perception of the RS pattern. We then present recent results in which eye movements serve as a tool for investigating perception-related brain states under general anaesthesia.

Disambiguating Stimulus Manipulations

In soliciting subjective responses from a monkey with stimuli that are either inherently ambiguous in nature, or near the perceptual threshold, it is essential to include the frequent presentation of control stimuli whose correct answer is clear to both the monkey and the experimenter. These so-called 'catch trials' provide a coarse measure of the monkey's performance, and comprise an essential part of nearly all studies in which the neural basis of perception is studied. In binocular rivalry, where monkeys are trained to report which of two stimuli they see at each point in time, unambiguous catch trials have been used in all studies to date (Leopold & Logothetis, 1996; Logothetis & Schall, 1989; Sheinberg & Logothetis, 1997). These generally consist of presenting the monocular stimuli alone, leading to an unambiguous response requirement, in the context of which inaccurate responses can be immediately identified. As a matter of course, provisions must be made to hide the identity of catch trials from the monkey, in our case by mimicking the smooth transitions of normal rivalry surrounding such trials. Such stimuli not only provide a means to monitor the performance, but also permit the experimenter to reward objectively based on response accuracy, and thus initially train and later continually remind the monkey not to stray from

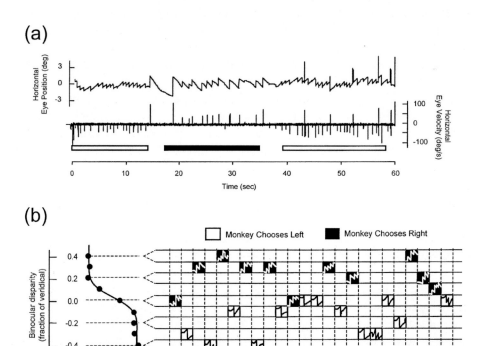

Figure 3

Behavioural controls used during monkey's viewing of the ambiguous rotating sphere shown in Figure 1. (a) Use of optokinetic nystagmus (OKN) responses as an objective indicator of perception. Eye position was measured during a one-minute period in which the monkey was viewing the ambiguous sphere. The upper trace is the horizontal eye position as a function of time, showing the characteristic sawtooth pattern of optokinetic movements. The change in the polarity of this pattern represents a change in the perceived spin direction of the rotating sphere. Immediately below is the eye velocity signal from the same period, whose vertical peaks (corresponding to the 'fast phase' of the OKN) show more clearly the perception-related state changes. Note that the change in the eye movements is not accompanied by a change in the stimulus, only a change in the perception. (b) Combination of several approaches used to evaluate response accuracy. In this experiment the monkey viewed brief periods of the rotating sphere, and was asked to report its direction each time by pulling one of two levers. To evaluate the response accuracy we used each of the approaches outlined in Figure 1. First, we applied a *stimulus manipulation* to bias the monkey's perception. We did this by applying binocular disparity (stereoscopic depth information) to the dots on the sphere. For large values of disparity, this disambiguated the percept sufficiently such that we could define a 'correct answer' for each trial. Second, we examined the *performance statistics* for disparity values that only slightly biased perception. This is shown in the left portion of the figure, where the fraction of leftward responses varies systematically as a function of the binocular disparity applied. Note that even with relatively small fractions of the 'correct' disparity (±0.2), the perception was systematically biased. Finally, the accuracy of performance was additionally evaluated by comparing the *objective physiological indicator* of OKN during epochs in which the monkey reported leftward vs. rightward motion. This is shown for a representative sample of 24 subsequently performed trials on the right side of the figure. On each trial, the sphere was shown with a particular disparity, and the monkey's response and OKN were recorded. Note that the responses and OKN were both influenced by the disparity applied, but that, under conditions of true ambiguity (0.0 disparity), the OKN polarity continued to match the monkey's report.

reporting his subjective percept. As described in greater detail below, monkeys are notoriously cagey in this respect. If they discover that there are no consequences for incorrect responses, their performance will tend toward completely random.

For other stimuli, such as the RS shown in Figure 2, effective catch trials can often be created by the addition of disambiguating cues. In the RS, the perception of three-dimensional shape can be traced to the parallax motion of dots, sliding past each other more quickly in the centre than near the edges. However, as in the Necker cube, there remain two equally valid interpretations, with the ball moving either clockwise or counter-clockwise when considered from above. This seldom occurs in natural vision, as the structure of any such pattern is fortified by redundant cues for depth, including binocular disparity,[1] shading, texture gradients, perspective, occlusion, etc. (Cavanagh, 1987). Experimentally, it is possible to use these and other cues to bias perception of the sphere with any one of these cues to create an unambiguous stimulus that can be employed as a catch trial. In the laboratory, we have used the cues of binocular disparity and luminance gradients to bias perception of the RS in this manner.

An example using binocular disparity to this end is shown on the right side of Figure 3b. In this experiment, perception of one or the other direction of the rotating sphere was enhanced by adding binocular information, and the amount of positive or negative disparity is signified by the vertical position in the plot. Black boxes signify a monkey reporting a rightward moving front face, and white boxes a leftward moving front face (for the moment, the eye movement traces inside the boxes should be ignored). Note that positive disparities (upper portion of the grid) yield primarily rightward responses while negative disparities (lower portion) yield primarily leftward responses. The addition of significant disparity is thus an effective cue to bias perception of the RS stimulus, providing us with a valid catch trial stimulus with which to continually evaluate and refine the accuracy of the monkey's perceptual report. This approach has been recently applied in neurophysiological experiments in monkeys trained to report the direction of rotation (Bradley *et al.*, 1998; Grunewald *et al.*, 2002; Dodd *et al.*, 2001).

The addition of external cues to disambiguate perception is not used exclusively to hone behavioural accuracy. It has also been used effectively to create alternate perceptual contexts for a given stimulus (Duncan *et al.*, 2000; Zhou *et al.*, 2000; Baumann *et al.*, 1997; Rossi *et al.*, 1996; Bakin *et al.*, 2000; Andrews *et al.*, 2002; for a review, see Albright & Stoner, 2002). These experiments generally rely on global stimulus changes that alter the manner in which an ambiguous local feature is interpreted. While such stimulus manipulations do not allow for a complete isolation of perceptual mechanisms, they nonetheless provide a powerful means to investigate how global information shapes perception-related responses in the brain.

---

[1] Binocular disparity refers to the slight shift in the retinal projections on the two eyes of a point in three-dimensional space. This shift stems from the slightly different views that the two eyes have of a target, and serves as the basis for our stereoscopic perception. In the laboratory, such disparity cues can be manipulated to add different amounts of stereoscopic depth information to a stimulus.

In some cases, the perception of a pattern can be biased not only by changes in its overall structure, but also by the temporal sequence of stimulus presentation. A series of relatively new tricks have made it possible to gain further experimental control over the perception of a stimulus that is, in many respects, completely ambiguous. An example of such a bias that has been used extensively in neurophysiological experiments is the phenomenon of binocular rivalry flash suppression (Wolfe, 1984). In this paradigm, dissimilar stimuli are presented to the two eyes asynchronously — with an interval >= 500 ms between the first and second. This presentation sequence consistently results in perception being dominated by the second pattern, with complete perceptual suppression of the first (until spontaneous binocular rivalry alternation sets in). Given this experimental control over perception, the flash suppression paradigm has been used in several neurophysiological studies to bias perception during binocular rivalry (Sengpiel et al., 1995; Sheinberg & Logothetis, 1997; Kreiman et al., 2002; Leopold et al., 2003). Interestingly, a recent study has suggested that flash suppression is not restricted to conditions of interocular conflict, but can be used effectively even if there is no spatial conflict between the first and second stimuli (Wilke et al., 2002). This 'generalized' flash suppression is a powerful new technique that is currently being used to investigate the neural basis of stimulus visibility. Importantly, and in contrast to other disappearance phenomena, both types of flash suppression are effective in the face of even the strongest attention or deepest introspection.

Finally, a recently described technique provides an additional means of having control over a monkey's percept during neurophysiological experiments. When ambiguous patterns are presented intermittently rather than continuously, perception can often become locked in a particular configuration for extended periods of time (Leopold et al., 2002b). In the context of this stabilization, the subjective interpretation of successive presentations is not akin to tossing a coin, but is instead highly dependent upon the preceding interpretations. This differs markedly from continuous viewing of ambiguous patterns, where the perceptual history is thought to have little if any role in dictating the stimulus interpretation at any point in time. The degree of stabilization is a variable of the exact stimulus timing parameters, thus providing an additional means to evaluate the accuracy of the monkey's perceptual responses. However, perhaps more importantly, this paradigm allows for the investigation of activity *between* stimulus presentations, where the perceptual interpretation of the upcoming presentation is, in fact, already determined (Maier et al., 2002). Stabilization, like flash suppression, can be observed under a wide range of attentional states, and is experienced equally well by naïve subjects (who are unaware the stimulus is ambiguous) and experienced subjects.

Many of the above stimulus manipulations are powerful in biasing perception, and are critical to the arsenal of tools used by neurophysiologists seeking fundamental mechanisms of vision. However, alone they are often insufficient to draw conclusions regarding the neural underpinnings of subjective perception for at least two reasons. First, from a physiological perspective, such perception-related changes are, by definition, confounded with a change in the visual stimulus. Often these changes need to be relatively large in order to provide an unambiguous interpretation for the monkey. The burden of proof lies with the investigator that perception-related modulation of neural activity is above and beyond that which might be expected based only on the sensory manipulation. Second, and more important for the current discussion, stimulus manipulations alone, in the form of catch trials, are often insufficient for enforcing accurate responses during truly ambiguous trials. This is because monkeys can often detect subtle changes in a stimulus, and may use that, rather than the intended global percept, to perform the task. The intelligence of monkeys is seldom clearer than when they find a shortcut or unintended cue to use in optimizing their reward. Their ability to discriminate ambiguous (uncontrolled) from unambiguous (controlled) trials is a particularly nefarious example of this. A monkey might, for example, modify his behaviour according to the ability of the experimenter to reward the accuracy of his response, reporting perfectly on unambiguous catch trials and randomly during the truly ambiguous trials (during which they are inevitably rewarded by chance, even if they report their percept inaccurately). While this might seem unlikely, monkeys commonly show such refined strategies in attempting to optimize their frequency of reward.

There are a number of commonly applied tricks to minimize the probability of the monkey making the association between stimulus appearance (unambiguous vs. ambiguous) and reward contingency (accurate feedback vs. no feedback). For example, one strategy, which is beyond the scope of the current discussion, is to employ a reward schedule that is variable, where the monkey always has a certain *probability* of obtaining reward with each correct response (a so-called variable-ratio schedule). Another trick, described below, is to restrict the disambiguating cues to low values, serving only to push perception of an ambiguous stimulus in a particular direction, without guaranteeing that it will follow 100% of the time. With such stimuli, it is often possible to create a biased stimulus that is nearly or even completely indiscriminable from the ambiguous pattern. But given that its bias is relatively weak, its effectiveness cannot be adequately judged on a trial-by-trial basis, but must rely on a statistical analysis. The next section thus focuses on how the performance and other statistics, when considered over many trials, can be used as evidence that a monkey is responding truthfully according to his subjective percept.

Statistical Analysis of Perception

A slight bias in an ambiguous pattern essentially affects the probability that it will be perceived in a particular configuration. The reason is that, while the stimulus is still lacking a large enough amount of disambiguating cues to be perceived in a unique fashion, the visual system can take into consideration even very small cues that favour one of the alternate stimulus interpretations. This happens more often the more salient these cues are, thus providing the possibility to parametrically vary the likelihood of a perceptual bias to occur. To return to the RS example, the left portion of Figure 3b shows that varying the strength of a cue along a continuum serves to change this probability along a continuum. In this figure, disparity values of +1.0 and −1.0 can be seen to correspond to the 'correct' binocular disparity that would exist in an actual three-dimensional rotating sphere of the size depicted on the screen. Notice that, in the well-trained monkey from whom these data were derived, even a relatively small disparity value (±0.2) was sufficient to impose a particular perceptual interpretation. Smaller disparities gave rise to intermediate levels of bias, with a value of 0.0 (no valid stereoscopic information) resulting in half the choices leftward and half rightward. Such performance statistics, giving rise to a psychometric function, have been used previously to demonstrate that the monkey is accurately reporting his perception of ambiguous or threshold-level stimuli (see, for example, Britten et al., 1992; Dodd et al., 2001). In the case of ambiguous or rivalrous patterns, it is important to acknowledge that there is no *a priori* reason to expect equal dominance of all competing percepts, since there are also subject-specific inherent biases in the manner in which stimuli are perceived (Sereno & Sereno, 1999).

Performance, in the sense of correct or 'correctly biased' responses, is but one statistical variable that can be monitored to establish the accuracy of the animal's report. In addition, one can monitor aspects of the temporal dynamics, a technique that has proven invaluable in the study of binocular rivalry. In rivalry the alternation between dominance of each eye's pattern is stochastic in time (Fox & Herrmann, 1967), with the duration of a particular dominance phase independent of the alternation history. However, despite the randomness in this process, the shape of the distribution of many dominance phases is entirely deterministic, and often modelled with a gamma function (Levelt, 1965). While the processes underlying these dynamics remain a topic of debate, they serve as a signature for rivalry and other forms of multistable perception (Borsellino et al., 1972; Leopold & Logothetis, 1999). These dynamics also characterize monkeys' perception during rivalry (Myerson et al., 1981; Leopold & Logothetis, 1996), and thus provide additional, although arduous, means to evaluate response accuracy during neurophysiological experiments. The mean or expected dominance time is also a parameter that can be manipulated in rivalry by adjusting the relative strength (e.g., contrast) of the competing monocular stimuli. This causes a stereotypic change in the relative balance of perception that was first described by Levelt (1965). Specifically, raising the contrast of one stimulus results in decreased mean dominance time of its rival. This result, besides raising the interesting question of why such a relationship might emerge, has been instrumental in neurophysiological studies of binocular rivalry (Leopold & Logothetis, 1996; Sheinberg & Logothetis, 1997), and has provided a powerful means of evaluating performance.

The use of statistical evaluation, in combination with stimulus manipulations, thus provides a set of tools by which one can systematically manipulate a stimulus, and compare the changes over many trials according to a set of well-defined expectations. In addition to performance, and the temporal dynamics mentioned above, one can also observe trends in other variables such as manual reaction times for additional evidence that the monkey is responding correctly. Of course, none of these methods can effectively be used to monitor trial-by-trial responses during truly ambiguous presentations, for which an objective 'lie detector' would be invaluable. The next section demonstrates that such behavioural controls are actually possible in some instances, and that their use in the study of perception may play an increasingly prominent role.

Objective Physiologic Indicators of Perception

Sometimes the body betrays subjective states, a fact that serves as the basis for the polygraph or lie detector. Visual perception has a small number of objective physiologic indicators. The dilation of the pupil has, for example, been shown to change in accordance with periods of dominance and perception during binocular rivalry (Lowe & Ogle, 1966; Richards, 1966). Clearly, any such measures would be of great value in assessing and correcting the performance of a monkey who has little inherent motivation for responding accurately and 'honestly'. By far the most powerful type of objective indicator of perception used today is the pattern of eye movements automatically elicited upon viewing certain types of visual patterns. Of the different types of eye movements that might be exploited for this purpose, it is, perhaps ironically, an evolutionarily ancient system that has proven the most useful for gauging subjective perception — optokinetic nystagmus or OKN. These movements are elicited when a subject attempts to hold a steady gaze while viewing a moving surface. The eyes are typically forced into a sequence of drifts and jumps that, when viewed as a function of time, have a stereotypic sawtooth pattern (see, for example, Figure 3a). This well-studied phenomenon is robust, and can be elicited by many types of moving stimuli under a variety of conditions. Its reflexive nature has previously been exploited to evaluate, for example, aspects of perception in infants too young to report their percept (Manny & Fern, 1990; Fox et al., 1979).

We have used this technique extensively with both BR and RS stimuli. Unlike the infant studies, our use of OKN has been to reinforce or refute a particular response provided by the monkey. Importantly, this is possible because not only can OKN reflect the direction of a moving visual pattern, but it can also accurately reflect the perceived direction of motion in an ambiguous stimulus. This has been shown previously for binocular rivalry in both humans (Enoksson, 1968) and monkeys (Logothetis & Schall, 1990). If, for example, rivalry is initiated between an upward- and downward-moving pattern, perception of the upward pattern will be accompanied by 'upward' OKN (with the slow phase moving upward). When, however, the perception switches to downward (with no change in the stimulus) so does the OKN polarity. This obligatory coupling between the direction of OKN and perception has provided an objective indicator of perception in a number of human rivalry studies (Fox et al., 1975; Leopold et al., 1995) and has also served as the basis for training and evaluating the accuracy of responses in neurophysiological studies (Logothetis & Schall, 1989).

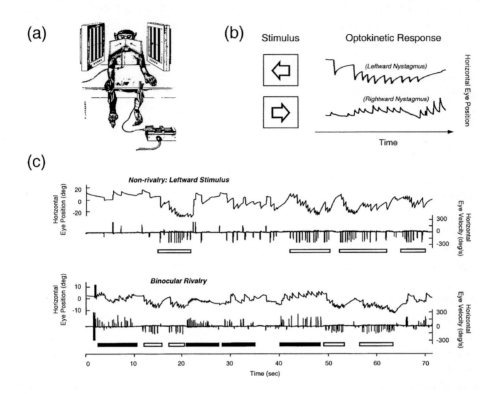

Figure 4

Perception-related optokinetic movements measured in a monkey in a state of dissociation anaesthesia brought about by the narcotic ketamine. (a) Setup designed to monitor physiological responses, including eye movements, during the gradual transition from wakefulness to anaesthesia. The setup is equipped with many features traditionally used to test alert monkeys, such as measurement of behavioural responses and eye position. In addition, the monkey has been trained to accept the insertion of an intravenous injection catheter during wakefulness, through which various agents, ketamine in this case, can be infused during the experiment. The visual stimulation apparatus is designed to present large-field motion patterns, with mirrors used to independently stimulate the two eyes. It was thus possible to present monocular, binocular, or binocularly rivalrous stimuli during both waking and anaesthesia. The dissociation anaesthesia brought about with ketamine results in the eyes remaining open, with the direction of gaze roughly straight ahead, permitting large-field stimulation with this apparatus. (b) Under low dose (1 mg/kg) ketamine anaesthesia, while the monkey is behaviourally dissociated from the environment, optokinetic movements can be elicited by moving stimuli, shown here by the different polarity traces for leftward vs. rightward stimulation. These OKN responses differ slightly in their structure from those elicited during waking, however, their polarity unambiguously reflects the direction of the inducing stimulus. (c) Examples of OKN patterns elicited during extended viewing of non-rivalrous and rivalrous moving patterns during ketamine dissociation. The format is the same as in Figure 2a. During non-rivalrous leftward stimulation, several trains of OKN movements were elicited with a polarity reflecting the stimulus (white bars). During binocular rivalry, below, the eye trace again shows trains of repetitive OKN movements induced by a stimulus. However, in this case, the leftward stimulus was presented to one eye while the rightward one was presented to the other one. In the waking state, this stimulus configuration normally gives rise to vigorous binocular rivalry, with perception continually alternating between rightward and leftward motion. As one can see from the figure, state changes in the OKN response were also present during ketamine anaesthesia. Given that the monkey was under a state of behavioural dissociation, and therefore unable to report any percept, it is interesting to speculate on the relationship of these state changes to conscious visual perception.

Recently, we have demonstrated that OKN is also a reliable indicator of perceived rotation for patterns like that shown in Figure 2a. An example of OKN movements during one minute of continuous, unrestricted viewing of the ambiguous rotating sphere is shown in Figure 3a. The top trace in this figure shows the horizontal eye position as a function of time. Note the characteristic sawtooth pattern, comprised of slow and fast phases, that is present throughout the trial. Close examination reveals that this pattern changed polarity after roughly 15 seconds, and then back again after roughly 38 seconds. These changes in polarity are more easily seen in the trace just underneath, which represents the instantaneous velocity at each time point. The fast phases (saccades) are shown by the spikes in this trace, with their sign showing a transition at the same time points. These phases of leftward and rightward OKN, further marked by the horizontal bars beneath, represent spontaneous state changes experienced by the monkey upon viewing this stimulus. These results are further supported by human psychophysical studies with superimposed patterns that are transparently moving in opposite directions. For these stimuli that are eliciting similar reversals of subjective depth, the evoked OKN were highly correlated with the perceived motion on the attended surface (Watanabe, 1999).

The eye movement traces shown in the grid on the right side of Figure 3b demonstrate that these optokinetic responses are in perfect agreement with the monkey's reported perception of the rotating sphere, just as is the case for binocular rivalry. Superimposed on the black squares, corresponding to trials in which the monkey pulled the lever signalling that the front face was perceived to the right, the OKN slow phases are moving rightward. On the white squares, however, the slow phases are moving leftward. This relationship is consistent even in the completely ambiguous condition (0.0 disparity). Thus, Figure 3b represents the convergence of each of the general techniques outlined in Figure 2, integrated to optimally control and evaluate the monkey's performance. Disambiguating cues were added to the pattern, a statistical analysis of performance was undertaken to verify the effect of subtle stimulus changes, and eye movements were monitored as an objective indicator of the monkey's perception.

Most recently, OKN has been used as an objective measure in monkeys whose perceptual state is not clearly defined (Figure 4). In these experiments, we investigated the possibility that optokinetic nystagmus could be used to investigate perception-related brain states in an animal in a state of 'dissociation' brought about by the anaesthetic ketamine (Leopold et al., 2002a). This experiment was facilitated by a novel setup in which the state of consciousness could be continually modulated, by injecting ketamine through an intravenous line (Figure 4a). The animal sat in a chair and viewed two mirrors, angled in such a way to provide independent stimulation to the two eyes from flat monitors. For a range of doses, OKN movements could be reliably elicited, albeit slightly abnormal in form (Figure 4b), while the monkey was otherwise unresponsive to external stimuli. Under these conditions, we presented either non-rivalrous, monocular stimulation (Figure 4c, upper) or rivalrous stimulation, with opposite horizontal directions in the two eyes (Figure 4c, lower). Interestingly, during binocular rivalry, we observed periods of leftward OKN interleaved with those of rightward OKN. During wakefulness such periods would correlate nearly perfectly with the monkey's perceptual experience. During ketamine dissociation, it is only possible to say that these OKN traces are objective indicators for brain-state changes that may bear a relationship to subjective perception. Thus, while it is interesting to speculate what such state changes might entail, further research is required to delve deeper into this difficult issue.

It is possible that objective indicators for perception will play an increasingly important role in brain research in the future, as our understanding of the brain's approach to vision becomes more refined. This might even take the form of using neural signals themselves as a means to 'decode' perception at each time point, and then using this signal to judge the accuracy of a subjective response. Physiological studies from our group and others have shown that, while many neurons in the early visual areas modulate with perception during ambiguous stimulation, these modulations are seldom reliable enough to serve as a behavioural control (for a review, see Leopold & Logothetis, 1999). Nonetheless, recordings from the temporal lobe visual areas suggest that with certain high-level neurons, such a scheme for monitoring perception may, in fact, be possible. During binocular rivalry flash suppression, neurons in monkeys, and more recently those in humans, reliably follow the subjective percept (Sheinberg & Logothetis, 1997; Kreiman et al., 2002). Remarkably, some neurons in the human medial temporal lobe even respond when a stimulus is not physically presented at all, but only *imagined* using visual imagery (Kreiman et al., 2000). The activity of these neurons thus appear to lie very close to very personal aspects of subjective perception, and experiments in the future may thus take an increasingly voyeuristic approach to exploring the neural underpinnings of visual perception.

Conclusions

The study of the neural basis of visual perception is a topic of great interest, and, while even its core remains poorly understood, considerable progress has been made. Ambiguous and rivalrous patterns serve as excellent tools for investigating principles of visual perception, and have been widely known and utilized long before the advent of alert monkey neurophysiology (von Helmholtz, 1925; Wertheimer, 1912; Koffka, 1935). Given their multiple perceptual interpretations, such patterns are a convenient tool for neurophysiologists seeking the neural basis of visual perception. They are diverse, appealing to many aspects of visual processing, easy to create, and a fascination for both visual scientists and the layman. One might then ask, why have there been so few studies that have specifically employed such patterns to study neural mechanisms of perceptual organization? The answer most certainly lies not in the difficulty of the physiological techniques, nor in the stimuli themselves, but rather in the great difficulty in obtaining a reliable measure of subjective visual perception from monkeys. The various tools outlined and elaborated here provide a concerted approach that we have used successfully to tackle this challenging problem. Further elaboration of these techniques, as well as the introduction of new ones, will likely hold an important place for neurophysiologists eager to understand how that which we sense is transformed into that which we see.

Acknowledgements

The authors would like to thank Margaret Sereno for providing the structure-from-motion stimuli, and Joachim Werner for technical assistance. This work was supported by the Max Planck Society.

References

Albright, T.D., Stoner, G.R. (2002), 'Contextual influences on visual processing', *Annual Review of Neuroscience*, **25**, pp. 339–79.
Andrews, T.J., Schluppeck, D., Homfray, D., Blakemore, C., Blakemore, C. (2002), 'Activity in the fusiform gyrus predicts conscious perception of Rubin's vase–face illusion', *Neuroimage*, **17** (2), pp. 890–901.
Bakin, J.S., Nakayama, K., Gilbert, C.D. (2000), 'Visual responses in monkey areas V1 and V2 to three-dimensional surface configurations', *Journal of Neuroscience*, 20 (21), pp. 8188–98.
Baumann, R., van der Zwan, R., Peterhans, E. (1997), 'Figure–ground segregation at contours: a neural mechanism in the visual cortex of the alert monkey', *European Journal of Neuroscience*, **9** (6), pp. 1290–303.
Borsellino, A., De Marco, A., Allazetta, A., Rinesi, S., Bartolini, B. (1972), 'Reversal time distribution in the perception of visual ambiguous stimuli', *Kybernetik*, **10** (3), pp. 139–44.
Bradley, D.C., Chang, G.C., Andersen, R.A. (1998), 'Encoding of three-dimensional structure-from-motion by primate area MT neurons', *Nature*, **392** (6677), pp. 714–7.
Britten, K.H., Shadlen, M.N., Newsome, W.T., Movshon, J.A. (1992), 'The analysis of visual motion: a comparison of neuronal and psychophysical performance', *Journal of Neuroscience*, **12** (12), pp. 4745–65.
Cavanagh, P. (1987), 'Reconstructing the third dimension: interactions between color, texture, motion, binocular disparity and shape', *Computer Vision, Graphics, and Image Processing*, 37, pp. 171–95.
Crick, F., Koch, C. (2003), 'A framework for consciousness', *Nature Neuroscience*, **6** (2), pp. 119–26.
Dodd, J.V., Krug, K., Cumming, B.G., Parker, A.J. (2001), 'Perceptually bistable three-dimensional figures evoke high choice probabilities in cortical area MT', *Journal of Neuroscience*, **21** (13), pp. 4809–21.
Duncan, R.O., Albright, T.D., Stoner, G.R. (2000), 'Occlusion and the interpretation of visual motion: perceptual and neuronal effects of context', *Journal of Neuroscience*, **20** (15), pp. 5885–97.
Dutour, E.F. (1760), 'Discussion d'une question d'optique [Discussion on a question of optics]', *l'Academie des Sciences. Memoires de Mathematique et de physique presentes par Divers Savants*, 3, pp. 514–30.
Enoksson, P. (1968), 'Studies in optokinetic binocular rivalry with a new device', *Acta Ophthalmologica*, **46** (1), pp. 71–4.
Fox, R., Herrmann, J. (1967), 'Stochastic properties of binocular rivalry alternations', *Perception & Psychophysics*, 2, pp. 432–6.
Fox, R., Aslin, R.N., Shea, S.L., Dumais, S.T. (1979), 'Stereopsis in human infants', *Science*, **207**, pp. 323–4.
Fox, R., Todd, S., Bettinger, L.A. (1975), 'Optokinetic nystagmus as an objective indicator of binocular rivalry', *Vision Research*, **15** (7), pp. 849–53.
Grunewald, A., Bradley, D.C., Andersen, R.A. (2002), 'Neural correlates of structure-from-motion perception in macaque V1 and MT', *Journal of Neuroscience*, **22** (14), pp. 6195–207.
Koffka, K. (1935), *Principles of Gestalt Psychology* (New York, Harcourt, Brace and World).
Kreiman, G., Fried, I., Koch, C. (2002), 'Single-neuron correlates of subjective vision in the human medial temporal lobe', *Proceedings of the National Academy of Sciences of the United States of America*, **99** (12), pp. 8378–83.
Kreiman, G., Koch, C., Fried, I. (2000), 'Imagery neurons in the human brain', *Nature*, **408** (6810), pp. 357–61.
Leopold, D.A., Fitzgibbons, J.C., Logothetis, N.K. (1995), 'The role of attention in binocular rivalry as revealed through optokinetic nystagmus', A.I.Memo No.1554, C.B.C.L.Paper No.126 , pp. 1–17.
Leopold, D.A., Logothetis, N.K. (1996), 'Activity changes in early visual cortex reflect monkeys' percepts during binocular rivalry', *Nature*, **379** (6565), pp. 549–53.
Leopold, D.A., Logothetis, N.K. (1999), 'Multistable phenomena: changing views in perception', *Trends in Cognitive Sciences*, **3** (7), pp. 254–64.

Leopold, D.A., Maier, A., Wilke, M., Logothetis, N.K. (2003), 'Binocular rivalry and the illusion of monocular vision' in *Binocular Rivalry and Perceptual Ambiguity*, eds D. Alais and R. Blake (Cambridge, MA: MIT Press).

Leopold, D.A., Plettenberg, H.K., Logothetis, N.K. (2002a), 'Visual processing in the ketamine-anesthetized monkey: optokinetic and blood oxygenation level-dependent responses', *Experimental Brain Research*, **143**, pp. 359–72.

Leopold, D.A., Wilke, M., Maier, A., Logothetis, N.K. (2002b), Stable perception of visually ambiguous patterns', *Nature Neuroscience*, **5** (6), pp. 605–9.

Levelt, W. (1965), *On binocular rivalry* (Soesterberg, The Netherlands: Institute for Perception RVO-TNO).

Logothetis, N.K., Schall, J.D. (1989), Neuronal correlates of subjective visual perception.', *Science*, **245** (4919), pp. 761–3.

Logothetis, N.K., Schall, J.D. (1990), 'Binocular motion rivalry in macaque monkeys: eye dominance and tracking eye movements', *Vision Research*, **30** (10), pp. 1409–19.

Lowe, S.W., Oglc, K.N. (1966), 'Dynamics of the pupil during binocular rivalry', *Archives of Ophthalmology*, **75** (3), pp. 395–403.

Maier, A., Leopold, D.A., Logothetis, N.K. (2002), 'Neural activity during stable perception of ambiguous displays in monkey visual cortex', *Soc. Neurosci. Abstr*, **161**, p. 13.

Manny, R.E., Fern, K.D. (1990), 'Motion coherence in infants', *Vision Research*, **30** (9), pp. 1319–29.

Myerson, J., Miezin, F., Allman, J. (1981), 'Binocular rivalry in macaque monkeys and humans: A comparative study in perception', *Behav. Anal. Lett.* **1**, pp. 149–56.

Richards, W. (1966), 'Attenuation of the pupil response during binocular rivalry', *Vision Research*, **6** (3), pp. 239–40.

Rossi, A.F., Rittenhouse, C.D., Paradiso, M.A. (1996), 'The representation of brightness in primary visual cortex', *Science*, **273** (5278), pp. 1104–7.

Sengpiel, F., Blakemore, C., Harrad, R. (1995), 'Interocular suppression in the primary visual cortex: a possible neural basis of binocular rivalry', *Vision Research*, **35** (2), pp. 179–95.

Sereno, M.E., Sereno, M.I. (1999), '2-D center-surround effects on 3-D structure-from-motion', *Journal of Experimental Psychology: Human Perception & Performance*, **25** (6), pp. 1834–54.

Sheinberg, D.L., Logothetis, N.K. (1997), 'The role of temporal cortical areas in perceptual organization', *Proceedings of the National Academy of Sciences of the United States of America*, **94** (7), pp. 3408–13.

von Helmholtz, H. (1925), *Treatise on Physiological Optics* [1826] (Dover, New York, Southall, J.P.).

Wallach, H., O'Connell, D.N. (1953), 'The kinetic depth effect', *Journal of Experimental Psychology*, **45**, pp. 205–17.

Watanabe, K. (1999), 'Optokinetic nystagmus with spontaneous reversal of transparent motion perception, '*Experimental Brain Research*, **129** (1), pp. 156–60.

Wertheimer, M. (1912), 'Experimentelle Studien über das Sehen von Bewegung', *Zeitschrift für Psychologie mit Zeitschrift für angewandte Psychologie*, **61** (161), p. 265.

Wilke, M., Leopold, D.A., Logothetis, N.K. (2002), 'Flash suppression without interocular conflict', *Society for Neuroscience Abstracts*, **161**, p. 15.

Wolfe, J.M. (1984), 'Reversing ocular dominance and suppression in a single flash', *Vision Research*, **24** (5), pp. 471–8.

Zhou, H., Friedman, H.S., von der Heydt, R. (2000), 'Coding of border ownership in monkey visual cortex', *Journal of Neuroscience*, **20** (17), pp. 6594–611.

Timothy D. Wilson

Knowing When To Ask
Introspection and the Adaptive Unconscious

The introspective method has come under attack throughout the history of psychology, yet it is widely used today in virtually all areas of the field, often to good effect. At the same time indirect methods that do not rely on introspection are widely used, also to good effect. This conundrum is best understood in terms of models of nonconscious processing and the role of consciousness. People have access to many of their feelings and emotions, and develop rich narratives about themselves and their social worlds. These conscious states, accessible to introspective reports, are often good predictors of people's behaviour. There is also a pervasive adaptive unconscious that is inaccessible via introspection. When using introspective reports researchers should be clear about which kinds of mental states they are trying to measure.

The introspective method has come under repeated attack from the time psychology began as an empirical science to the present day (Jack and Roepstorff, 2002; Lieberman, 1979; Nisbett and Wilson, 1977), yet the method is alive and well in virtually all areas of the field. How can a method that is so widely used be so maligned? To understand this conundrum, we need to ask what people are doing when they introspect and what they are accessing. By developing clearer ideas about the limits of people's ability to access their cognitive and emotional processes we will know better what to ask people to report and what is better left unasked.

Introspective Reports are Alive and Well

A broad examination of psychological research — not just in cognitive psychology — reveals that researchers continue to put introspective reports to good use. A casual perusal of journals in virtually all subdisciplines of the field reveals many dependent measures in which participants are asked to report their internal states, and in each of these fields there is evidence for the validity of these reports. Consider these examples:

Personality research

Personality researchers have long relied on self-report inventories in which people rate their own traits. Some of these measures are quite direct and transparent, such as a recent measure of the 'big five' personality domains by Gosling, Rentfrow and Swann (2002), in which people rate their level of agreement with statements such as, 'I see myself as extraverted, enthusiastic' and 'I see myself as calm, emotionally stable'. The predictive validity of self-report measures of personality has been controversial for many years. No one doubts that there are correlations between such measures and actual behaviour, though there is much debate over the magnitude of these correlations (e.g. Epstein, 1979; Mischel, 1968; Ross and Nisbett, 1991). Recent evidence suggests that self-report measures are especially likely to predict behaviours that are under people's conscious control, as opposed to more spontaneous, uncontrollable behaviours (e.g. Asendorpf *et al.*, 2002).

Emotions research

In the thriving study of human emotion, the primary empirical method is the straightforward self-report questionnaire, on which people rate their current feelings, moods and emotional states. Many well-validated self-report instruments have been developed, such as the Multiple Affect Adjective Checklist (Zuckerman and Lubin, 1965) and the Positive Affect Negative Affect Schedule (Watson *et al.*, 1988). It is difficult to imagine the field of emotion research progressing as it has without the ability to question people about the nature of their feelings and emotions. Self-reported emotions have been found to predict many important behaviours, including prosocial acts, aggression, and suicide (e.g. Carlson *et al.*, 1988; Lindsay and Anderson, 2000; Nierenberg *et al.*, 2001; Sanchez and Le, 2001).

Attitudes research

Attitudes are one of the oldest topics in social psychology and the most common way of measuring them is with self-report questionnaires. In countless surveys and laboratory studies, people are asked to report their evaluations of social issues, consumer products, and other people, and these reports are, under many conditions, excellent predictors of behaviour. Ajzen and Fishbein's (1980) theory of planned behaviour, for example, argues that the best way to predict people's behaviour is to ask them their intentions, and such self-report measures have been found to be excellent predictors of behaviours such as people's use of birth control (e.g. Davidson and Jaccard, 1979). Similarly, political scientists have done a reasonable job of predicting the outcome of political elections with polls, in which people are asked for whom they plan to vote (e.g. Krosnick, 1999; Manza *et al.*, 2002; Miller, 2002). Simple questions about people's attitudes and intentions have considerable predictive validity, in that they often correlate highly with people's overt behaviour (e.g. Ajzen, 1996; Fazio, 1990).

Memory research

Since the inception of research on human memory in experimental psychology, one of the chief measurement techniques has been people's reports about what they recognize or recall. Had psychologists not relied on these kinds of self-reports, research on memory would never have started. Many current studies of memory continue to ask participants to report what they can recall (e.g. Smilek *et al.*, 2002) and the processes they used to recall something (e.g. Jacoby *et al.*, 1997). Several studies have found that these types of self-report measures predict interesting outcomes, such as people's accuracy in identifying suspects in line-ups (e.g. Dunning and Stern, 1994).

Developmental research

An active area of research in developmental psychology is the study of children's theory of mind and attributions about the causes of their own and other people's behaviour. Many of these studies rely heavily on children's reports about themselves and their environments, such as their beliefs about where someone will look for a hidden object (in the false belief paradigm; e.g. Wimmer and Perner, 1983) or their reports about other people's desires and attitudes (e.g. in the literature on discounting and attribution; Karniol and Ross, 1976). For example, a large literature has examined children's ability to predict other people's behaviour, such as where people will look for a hidden object. By age four or so children's predictions are quite accurate (e.g. Wellman *et al.*, 2001; Perner and Clements, 2000).

In each of these areas of research, the introspective method is alive and well and has yielded highly useful results; indeed, verbal reports are one of the main methods of the discipline. Why, then, all the fuss? Why have some, including me, questioned the use of the introspective method (Nisbett and Wilson, 1977; Wilson and Stone, 1985; Wilson, 1994)?

The Limits of Introspective Reports

Despite the heavy reliance on self-reports in each of the areas it is clear that the measures have their limits. One problem is the well-recognized issue of social desirability, namely whether people are willing to report the states that the researcher is interested in measuring.

A more fundamental problem is whether participants have access to their thoughts and feelings. Increasingly, researchers are realizing that a great deal of what they are interested in measuring is not consciously accessible to their participants, forcing them to rely on alternative methods. Indeed, in each of the areas I have just reviewed, the limits of people's ability to report their internal states and traits has been questioned:

Personality research

There is a long tradition of measuring personality with techniques other than verbal reports. In the early years of personality research projective techniques, such as the Rorschach and Thematic Apperception Test, were developed to measure aspects of personality that people might be repressing (e.g. Murray, 1938). In recent years there has been an increased emphasis on implicit measures of personality that do not rely on people's introspective access to their own personalities, and may tap states that are simply inaccessible to people (i.e. are not necessarily the result of repression but still cannot be accessed consciously; for a review, see Wilson, 2002).

Emotions research

Researchers have long sought physiological and behavioural measures of emotion, including facial expressions, observer ratings, and autonomic responses. One advantage of these measures is that they avoid the unwillingness people might have to report their true feelings. As in the area of personality research, theorists with a psychoanalytic bent have argued that people sometimes have difficulty reporting some of their feelings due to repression, though whether nonconscious feelings exist (or can be proven to exist) is controversial. With the advent of implicit measures and new theories about brain functioning, many researchers now argue that some feelings are inaccessible to conscious awareness, not necessarily because of repression, but because of the architecture of the mind (e.g. LeDoux, 1996; Wilson, 2002). Consequently, nonverbal measures of emotion are crucial, though as noted by Larsen and Fredrickson (1999), none have proven to be 'the gold standard' to which all other measures can be compared.

Attitudes research

The search for indirect measures of attitudes also has a long history (see, for example, Crosby, Bromley and Saxe, 1980). With the recent development of implicit measures such as the Implicit Associations Test (Greenwald, McGhee and Schwartz, 1998), there has been an explosion of research on the nature of attitudes and people's awareness of them. Some have argued that people can simultaneously possess different implicit and explicit attitudes towards the same object, with self-report measures tapping only the explicit attitude (e.g. Wilson, Lindsey and Schooler, 2000).

Memory research

Although verbal report measures have been indispensable to research on memory and information processing, it was the development of indirect measures (chiefly reaction time) that fuelled the cognitive revolution in the 1950s and 1960s (e.g. Teichner and Krebs, 1974; Wood and Jennings, 1976). Cognitive psychologists have also relied heavily on measures of brain activity, including electroencephalogram (EEG) recordings, positron emission tomography (PET) scans and functional magnetic resonance imaging (fMRI). Clearly these measures reveal workings of the brain to which participants have no access.

Developmental research

Although children's verbal reports are still the primary method in many areas of developmental psychology, questions about the limits of their access to their own thought processes have been raised, and alternative measures, such as gaze and actual behaviour, often reveal different developmental trajectories of cognitive processes (see Wilson, 2002, Ch. 3 for examples).

In each of these areas researchers are developing theories about the kinds of mental processes and states to which people have direct access (e.g. explicit attitudes, self-theories about personality) and the processes and states to which they do not (e.g. implicit attitudes and traits). Given that issues about the validity of self-reports have been percolating in psychology for decades, it is curious that theoretical progress about the nature of conscious versus unconscious processing has been so slow.

A Brief Historical Overview

The debate over the validity of introspection began at the onset of psychology as an empirical science in the late nineteenth and early twentieth centuries (e.g. Titchener, 1909; Wundt, 1894). At that time the main subject matter of psychology was 'the stuff of which consciousness was made', studied by presenting stimuli to participants and asking them to introspect 'on the course of the experience that has just taken place' (Allport, 1955, p. 71). This method turned out to be a dead end, due largely to the unreliability of reports from one laboratory to the next.

This early failure of the introspective method could, in principle, have led in many fruitful directions. On the one hand, psychologists might have concluded that the basic processes they were trying to study, the building blocks of sensation and perception, were inaccessible to conscious awareness. Such a conclusion could have been a stepping stone to hypotheses about the accessibility of mental processes more generally, leading to theories about the power and scope of nonconscious mental processes.

There were two major obstacles to such theoretical progress. First was the long shadow cast by psychoanalysis, the major theory of the unconscious at the time. The view of the unconscious as a dynamic set of instinctual urges dominated the theoretical stage, making it difficult to develop alternative theories of a pervasive, adaptive unconscious. Second, there few methods available to study nonconscious processes. Experimental psychologists willing to tackle such difficult-to-test notions as repression and perceptual defence were hampered by inadequate methods (for reviews see Erdelyi, 1974, 1985). Research on perceptual defence, for example, became bogged down in thorny methodological issues and largely petered out until new methods were developed to test the influence of subliminal exposures to stimuli (see, for example, Dijksterhuis and Bargh, 2001). There simply were no sound methods available to test hypotheses about unconscious processes.

Though psychoanalysis was the dominant theory about unconscious processes, ironically there were prior formulations to which researchers frustrated by the failure of the introspective method might have turned. A generation before Freud, a group of nineteenth-century British physicians and philosophers argued for the existence of a different kind of unconscious, one that bore a remarkable similarity to the current view of a set of pervasive, adaptive, sophisticated, mental processes that are critical to human survival but which operate in parallel to consciousness. Laycock (1860), Hamilton (1865) and Carpenter (1874) pointed to the existence of powerful nonconscious processes that analyse the world and direct behaviour automatically, and are not just the repository of threatening, infantile urges that are pushed out of awareness, as Freud was to argue shortly (see Miller, 1995; Wilson, 2002).

Assuming that the early experimental psychologists were aware of these views, and were willing to consider that the failure of their introspective methods might be evidence for them, they would still have been stymied by the absence of tools to test ideas about a powerful set of unconscious mental processes. As with any science, theory and methodology exist in a symbiotic relationship in psychology, and the absence of appropriate methods hindered theoretical developments about nonconscious processing.

Instead of leading to theoretical developments about the nature of unconscious mental processes, the failure of the introspective method led to theoretical developments in the opposite direction, away from theories about the mind (conscious or unconscious) and into the lap of behaviourism (e.g. Lieberman, 1979; Watson, 1913). Again, methodology drove theory: because there were no reliable ways of studying the mind, it was, according to behaviourists like Watson, unworthy of study, leading to several decades of research on the behaviour of pigeons and rats.

Credit for the revival of mentalism in psychology is often attributed to the cognitive revolution of the 1950s, indeed, experimental psychologists did return with vigour to the study of human information processing. One reason for this revival was the development of new methods that could illuminate mental processing with precision, without relying too heavily on introspective reports (chiefly reaction time). It is worthy of note, however, that the study of the human mind was alive and well in another area of psychology, before the cognitive revolution. Social psychologists never abandoned the experimental study of such mental processes and states as attitudes, persuasion, frames of reference, schemas, and social perception (Allport, 1985; Jones, 1985; Zajonc, 1980).

Neither social nor cognitive psychology, however, had much to say about the extent to which the mental processes they were studying were conscious or unconscious. In all likelihood this was due to skittishness over the respectability of psychology as a science. Research psychologists had to defend themselves against behaviourists, arguing that the mind could be studied as scientifically as the bar-pressing behaviour of rats. To go a step further and argue that much of what they were studying was unconscious mental processes would have made them vulnerable to the criticism that their ideas were as untestable as psychoanalytic theory.

As the study of the mind flourished, however, questions about people's access to their own mental processes became unavoidable. Rather than ignoring the fact that their research participants seemed unable to report anything like the cognitive processes they were hypothesized to have, psychologists began to entertain ideas about the limits of conscious awareness (e.g. Mandler, 1975; Nisbett and Wilson, 1977). These questions were controversial, but with the development and refinement of methods to examine nonconscious processes, such as subliminal exposures of stimuli and implicit learning paradigms, questions about the limits of conscious awareness and the scope and power of nonconscious mental processing became commonplace (e.g. Bargh and Pietromonaco, 1982; Greenwald, Draine and Abrams, 1996; Kunst-Wilson and Zajonc, 1980; Reber, 1993).

The Adaptive Unconscious versus Conscious Processing

Much work needs to be done to elucidate the nature and scope of nonconscious processing and how it interacts with conscious processing. The outline of a picture has emerged, however, and can be summarized as follows (see Wilson, 2002 for a more complete discussion):

1. Consciousness is a small part of human functioning, even smaller than the 'tip of the iceberg' Freud imagined. There is a pervasive set of mental processes that can be referred to as an *adaptive unconscious*, adaptive in the sense that these processes are vital to human survival.
2. One view of the mind is that low-level processes such as those involved in perception are nonconscious, whereas higher-order thinking and information processing are reserved for consciousness. According to this view, consciousness is the chief executive of the mind, setting policy and making major decisions, while nonconscious modules carry out more mundane mental tasks. This view has become untenable, however, as more and more research (largely by social psychologists) demonstrates the power and scope of nonconscious processes. Bargh and Chartrand (1999), for example, argue persuasively that mental processes previously thought to be the function of consciousness, such as the implementation of goals and the evaluation of one's experiences, often occur nonconsciously. The exact terrain of what people cannot access consciously continues to be mapped, but as the lines are redrawn, more and more mental territory is being allocated to nonconsciousness — so much so that some have argued that consciousness is a vastly overrated commodity, and that our very sense of having consciously willed our actions is often an illusion (Wegner, 2002).
3. As demonstrated by the research reviewed earlier that relies on introspective reports, people's self-views are not epiphenomenal. People have access to many of their feelings and emotions and develop rich narratives about

themselves and their social worlds. These conscious states often play an important causal role in people's behaviour. The conscious narratives people tell about themselves, for example, can be critical to their psychological well being (see Wilson, 2002, Ch. 8).
4. The result is a dual system of nonconscious traits, attitudes and emotions that underlie conscious versions of these states, often quite different from their nonconscious counterparts. Progress is being made in fleshing out the precise nature of this duality (e.g. Nosek and Banaji, 2002), though much work needs to be done (Jack and Shallice, 2001).

The basic question of what people have access to and what they do not is of course critical to the issue of when researchers can rely on introspective methods. The answer will come from careful research showing the predictive validity of verbal reports versus other measures of the construct under study. This statement is hardly novel but takes on new urgency as measures of implicit states continue to be developed and pitted against explicit measures of the same construct. Introspective reports will never disappear as a valued tool in psychological research, though their limits are becoming ever clearer.

Acknowledgement

The writing of this article was supported by a grant from the National Institute of Mental Health (RO1-MH56075).

References

Ajzen, I. (1996), 'The directive influence of attitudes on behavior', in *The Psychology of Action: Linking Cognition and Motivation to Behavior*, ed. P.M. Gollwitzer and J.A. Bargh (New York: Guilford).
Ajzen, I. and Fishbein, M. (1980), *Understanding Attitudes and Predicting Social Behavior* (Englewood Cliffs, NJ: Prentice Hall).
Allport, F.H. (1955), *Theories of Perception and the Concept of Structure* (New York: John Wiley & Sons).
Allport, G.W. (1985), 'The historical background of social psychology', in *Handbook of Social Psychology*, ed. G. Lindzey and E. Aronson (3rd edn., New York: Random House), Vol. 1, pp. 1–46.
Asendorpf, J., Banse, R. and Mücke, D. (2002), 'Double dissociation between implicit and explicit personality self-concept: the case of shy behavior', *Journal of Personality and Social Psychology*, **83**, pp. 380–93.
Bargh, J.A. and Chartrand, T.L. (1999), 'The unbearable automaticity of being', *American Psychologist*, **54**, pp. 462–79.
Bargh, J.A. and Pietromonaco, P. (1982), 'Automatic information processing and social perception: the influence of trait information presented outside of conscious awareness on impression formation', *Journal of Personality and Social Psychology*, **43**, pp. 437–49.
Carlson, M., Charlin, V. and Miller, N. (1988), 'Positive mood and helping behavior: a test of six hypotheses', *Journal of Personality and Social Psychology*, **55**, pp. 211–29.
Carpenter, W.B. (1874), *Principles of Mental Physiology* (New York: D. Appleton & Co.).
Crosby, F., Bromley, S. and Saxe, L. (1980), 'Recent unobtrusive studies of black and white discrimination and prejudice: a literature review', *Psychological Bulletin*, **87**, pp. 546–63.
Davidson, A.R. and Jaccard, J.J. (1979), 'Variables that moderate the attitude-behavior relation: results of a longitudinal survey', *Journal of Personality and Social Psychology*, **37**, pp. 1364–76.

Dijksterhuis, A. and Bargh, J.A. (2001), 'The perception–behavior expressway: automatic effects of social perception on social behavior', in *Advances in Experimental Social Psychology*, ed. M.P. Zanna (San Diego, CA: Academic Press), Vol. 33, pp. 1–40.
Dunning, D. and Stern, L.B. (1994), 'Distinguishing accurate from inaccurate eyewitness identifications via inquiries about decision processes', *Journal of Personality and Social Psychology*, **67**, pp. 818–35.
Epstein, S. (1979), 'The stability of behavior: I. On predicting most of the people much of the time', *Journal of Personality and Social Psychology*, **37**, pp. 1097–126.
Erdelyi, M. (1974), 'A new look at the New Look: perceptual defense and vigilance', *Psychological Review*, **81**, pp. 1–25.
Erdelyi, M. (1985), *Psychoanalysis: Freud's Cognitive Psychology* (New York: Freeman).
Fazio, R.H. (1990), 'Multiple processes by which attitudes guide behavior: The MODE model as an integrative framework', in *Advances in Experimental Social Psychology*, ed. M.P. Zanna (San Diego, CA: Academic Press), Vol. 23, pp. 75–109.
Gosling, S.D., Rentfrow, P.J. and Swann, W.B., Jr. (2002), *Very Brief Measure of the Big-Five Personality Domains* (unpublished manuscript).
Greenwald, A.G., Draine, S.C., Abrams, R.L. (1996), 'Three cognitive markers of unconscious semantic activation', *Science*, **273**, pp. 1699–702.
Greenwald, A.G., McGhee, D.E. and Schwartz, J.L.K. (1998), 'Measuring individual differences in implicit cognition: the implicit association test', *Journal of Personality and Social Psychology*, **74**, pp. 1464–80.
Hamilton, W. (1865), *Lectures on Metaphysics*, Vol. 1 (Boston, MA: Gould and Lincoln).
Jack, A.I. and Roepstorff, A. (2002), 'Introspection and cognitive brain mapping: from stimulus-response to script-report', *Trends in Cognitive Sciences*, **6**, pp. 333–9.
Jack, A.I. and Shallice, T. (2001), 'Introspective physicalism as an approach to the science of consciousness', *Cognition*, **79**, pp. 161–96.
Jacoby, L.L., Yonelinas, A.P. and Jennings, J.M. (1997), 'The relation between conscious and unconscious (automatic) influences: a declaration of independence', in *Scientific Approaches to Consciousness*, ed. J.D. Cohen and J.W. Schooler (Mahwah, NJ: Lawrence Erlbaum), pp. 13–47.
Jones, E.E. (1985), 'Major developments in social psychology during the past five decades', in *Handbook of Social Psychology*, ed. G. Lindzey and E. Aronson (3rd edn., New York: Random House), Vol. 1, pp. 47–108.
Karniol, R. and Ross, M. (1976), 'The development of causal attributions in social perception', *Journal of Personality and Social Psychology*, **34**, pp. 455–64.
Krosnick, J.A. (1999), 'Survey research', *Annual Review of Psychology*, **50**, pp. 537–67.
Kunst-Wilson, W.R. and Zajonc, R.B. (1980), 'Affective discrimination of stimuli that cannot be recognized', *Science*, **207**, pp. 557–8.
Larsen, R.J. and Fredrickson, B.L. (1999), 'Measurement issues in emotion research', in *Well-Being: The Foundations of Hedonic Psychology*, ed. D. Kahneman, E. Diener and N. Schwarz (New York: Russell Sage Foundation).
Laycock, T. (1860), *Mind and Brain: The Correlations of Consciousness and Organization* (London: Simpkin, Marschall, & Co.).
LeDoux, J. (1996), *The Emotional Brain: The Mysterious Underpinnings of Emotional Life* (New York: Simon & Schuster).
Lieberman, D.A. (1979), 'Behaviourism and the mind: a (limited) call for a return to introspectionism', *American Psychologist*, **34**, pp. 319–33.
Lindsay, J.J. and Anderson, C.A. (2000), 'From antecedent conditions to violent actions: a general affective aggression model', *Personality and Social Psychology Bulletin*, **26**, pp. 533–47.
Mandler, G. (1975), 'Consciousness: respectable, useful, and probably necessary', in *Information Processing and Cognition: The Loyola Symposium*, ed. R. Solso (Hillsdale, NJ: Erlbaum).
Manza, J., Cook, F.L. and Page, B.I. (ed. 2002), *Navigating Public Opinion : Polls, Policy, and the Future of American Democracy* (New York : Oxford University Press).
Miller, J. (1995), 'Going unconscious', *New York Review of Books* (April 20), pp. 59–65.
Miller, P.V. (2002), 'The authority and limitation of polls', in *Navigating Public Opinion*, ed. J. Manza, F.L. Cook and B.I. Page (New York: Oxford University Press).
Mischel, W. (1968), *Personality and Assessment* (New York: Wiley).
Murray, H.A. (1938), *Explorations in Personality* (New York: Oxford University Press).

Nierenberg, A.A., Gray, S.M. and Grandin, L.D. (2001), 'Mood disorders and suicide', *Journal of Clinical Psychiatry*, **62**, pp. 27–30.

Nisbett, R.E. and Wilson, T.D. (1977), 'Telling more than we can know: verbal reports on mental processes', *Psychological Review*, **84**, pp. 231–59.

Nosek, B.A. and Banaji, M.R. (2002), '(At least) two factors moderate the relationship between implicit and explicit attitudes', in *Natura Automatyzmow*, ed. R.K. Ohme and M. Jarymowicz (Warszawa: WIP PAN and SWPS).

Perner, J. and Clements, W.A. (2000), 'From and implicit to an explicit "theory of mind" ', in *Beyond Dissociation: Interaction Between Dissociated Implicit and Explicit Processing*, ed. Y. Rossetti and A. Revonsuo (Amsterdam: John Benjamins).

Reber, A.S. (1993), *Implicit Learning and Tacit Knowledge: An Essay on the Cognitive Unconscious* (New York: Oxford University Press).

Ross, L. and Nisbett, R.E. (1991), *The Person and the Situation* (New York: McGraw-Hill).

Sanchez, L. and Le, L.T. (2001), 'Suicide in mood disorders', *Depression & Anxiety*, **14**, pp. 177–82.

Smilek, D., Dixon, M.J., Cudahy, C. and Merikle, P.M. (2002), 'Synesthetic color experiences influence memory', *Psychological Science*, **13**, pp. 548–52.

Teichner, W.H. and Krebs, M.J. (1974), 'Laws of visual choice reaction time', *Psychological Review*, **81**, pp. 75–98.

Titchener, E.B. (1909), *Experimental Psychology of the Thought Processes* (New York: Macmillan).

Watson, D., Clark, L.A. and Tellegen, A. (1988), 'Development and validation of brief measures of positive and negative affect: the PANAS scales', *Journal of Personality and Social Psychology*, **54**, pp. 1063–70.

Watson, J.B. (1913), 'Psychology as the behaviorist views it', *Psychological Review*, **20**, pp. 158–77.

Wegner, D.M. (2002), *The Illusion of Conscious Will* (Cambridge, MA: MIT/Bradford).

Wellman, H.M., Cross, D. and Watson, J. (2001), 'Meta-analysis of theory-of-mind development: the truth about false belief', *Child Development*, **72**, pp. 655–84.

Wilson, T.D. (1994), 'The proper protocol: validity and completeness of verbal reports', *Psychological Science*, **5**, pp. 249–52.

Wilson, T.D. (2002), *Strangers to Ourselves: Discovering the Adaptive Unconscious* (Cambridge, MA: Harvard University Press).

Wilson, T.D., Lindsey, S. and Schooler, T. (2000), 'A model of dual attitudes', *Psychological Review*, **107**, pp. 101–26.

Wilson, T.D. and Stone, J.I. (1985), 'Limitations of self-knowledge: more on telling more than we can know', in *Review of Personality and Social Psychology*, ed. P. Shaver (Beverly Hills, CA: Sage), Vol. 6, pp. 167–83.

Wimmer, H. and Perner, J. (1983), 'Beliefs about beliefs: representation and constraining function of wrong beliefs in young children's understanding of deception', *Cognition*, **13**, pp. 103–28.

Wood, C.C. and Jennings, J.R. (1976), 'Speed–accuracy tradeoff functions in choice reaction time: experimental designs and computational procedures', *Perception & Psychophysics*, **19**, pp. 92–101.

Wundt, W. (1894), *Lectures on Human and Animal Psychology*, trans. S.E. Creigton and E.B. Titchener (New York: Macmillan).

Zajonc, R.B. (1980), 'Cognition and social cognition: a historical perspective', in *Retrospections on Social Psychology*, ed. L. Festinger (New York: Oxford).

Zuckerman, M. and Lubin, B. (1965), *The Multiple Affect Adjective Checklist* (San Diego: Educational Testing Service).

Gualtiero Piccinini

Data from Introspective Reports
Upgrading from Common Sense to Science

Introspective reports are used as sources of information about other minds, in both everyday life and science. Many scientists and philosophers consider this practice unjustified, while others have made the untestable assumption that introspection is a truthful method of private observation. I argue that neither scepticism nor faith concerning introspective reports are warranted. As an alternative, I consider our everyday, commonsensical reliance on each other's introspective reports. When we hear people talk about their minds, we neither refuse to learn from nor blindly accept what they say. Sometimes we accept what we are told, other times we reject it, and still other times we take the report, revise it in light of what we believe, then accept the modified version. Whatever we do, we have (implicit) reasons for it. In developing a sound methodology for the scientific use of introspective reports, we can take our commonsense treatment of introspective reports and make it more explicit and rigorous. We can discover what to infer from introspective reports in a way similar to how we do it every day, but with extra knowledge, methodological care and precision. Sorting out the use of introspective reports as sources of data is going to be a painstaking, piecemeal task, but it promises to enhance our science of the mind and brain.

I: Introduction: To Introspect or Not to Introspect?

I *feel* tired, this page *looks* white to me, but I'm *thinking* I can't procrastinate any more — these are *introspective reports*. In everyday life, we rely on reports like these to learn about other minds. This paper discusses whether introspective reports are also legitimate sources of scientific evidence. Introspection is often associated with consciousness, and introspective reports may well be particularly useful in the study of consciousness. But the present topic is not consciousness or experience or qualia or any particular kind of mental state, as construed by a scientific theory or by folk psychology. I will not address the status of folk psychology or its relation with scientific psychology. My topic is the use of introspective reports to generate evidence. What that evidence is *about* is a separate question, on which I will remain as neutral as possible.

Science is supposed to be based on *public*, or intersubjective, methods. Public methods are such that (i) different investigators can apply them to answer the same questions, and (ii) when they do so — other things being equal — they obtain the same results (Piccinini, forthcoming). But introspective reports, when they are expressed sincerely,[1] are often construed as the output of a non-public method of observation: I introspect my mind not yours; you introspect your mind not mine. Under this construal, introspective reports violate (i) above: different investigators cannot answer the same questions about the same minds by introspecting.

Because they are private, introspective reports are sometimes said to be in principle *unverifiable*, in the sense that aside from the introspecting subject, no one is in a position to determine whether a report is true or false.

The *introspection agnostic* thinks that since introspection is private and introspective reports are consequently unverifiable, scientists shouldn't take a stance on their truth value. According to the agnostic, scientists should not treat introspective reports as a special source of information about mental states (Lyons, 1986; Dennett, 1991). Instead, scientists should treat introspective reports as observable behaviours, to be explained on a par with all other behaviours (Dennett, 1991; 2001).

The *introspection believer*, by contrast, argues that introspective reports provide otherwise unavailable evidence, which we could not collect by any other means. The believer agrees with the agnostic that introspection is private and introspective reports are unverifiable. Because of this, the believer admits that introspection's truthfulness cannot be established by public methods. Nevertheless, the believer encourages scientists to assume that introspective reports are true at least most of the time, and hence they are legitimate sources of scientific evidence (Goldman, 1997; 2000; Chalmers, 1996). According to the believer, scientists should treat introspection as a genuine method of observation, yielding a special sort of first-person (private) data about mental states. These first-person data are special in that they cannot be obtained by third-person (or public) methods. The result is a new kind of first-person science — radically different from ordinary third-person science (Goldman, 1997; 2000; Chalmers, 1996; 1999; Price & Aydede, submitted).

Neither of the above attitudes is satisfactory. The problem with the agnostic position is that introspective reports do seem to be a precious source of information about minds, which many scientists are eager to exploit in their research (Jack & Roepstorff, 2002). Rather than rejecting introspective reports as sources of evidence, it would be preferable to look for a sound way to learn from them. Learning from them is what the believer wants to do, but all she offers to underwrite introspection is the untestable assumption that it's truthful. This is more akin to *faith* than sound scientific methodology. If we are going to use introspective reports as sources of scientific data, we'd better have good reasons.

---

[1] Unless otherwise noted, from now on I will omit this qualification and talk about introspective reports that are expressed *sincerely*.

In this paper, I argue that neither agnosticism nor faith concerning introspective reports is warranted. As an alternative, I consider our everyday, commonsensical reliance on each other's introspective reports. When we hear people talk about their minds, we neither blindly accept what they say nor refuse to learn from it. Instead, we (implicitly) weigh introspective reports against two relevant bodies of evidence: our beliefs about people and their circumstances, and our beliefs about the specific person and circumstance that generated the reports. Sometimes we accept what we are told, other times we reject it or suspend judgement, and still other times we take the report, revise it in light of what we believe, then accept the modified version. For example, if my daughter tells me she's still hungry after a regular meal, I may reasonably infer that she wants attention, not food. Whether we accept, reject or revise introspective reports, we have (implicit) reasons for it.

And reasons can be made explicit. When we assess the accuracy of our neighbours' introspective reports, we do so by means that are (or can be made) public. If we — who have no scientific theories or methodology — can do this, so can scientists. A sound scientific use of introspective reports can be based on a more explicit and theoretically sophisticated version of the same methodology. The proper scientific use of introspective reports goes hand in hand with the requirement that scientific methods be public — as it should.

This leads me to reject an assumption that's common to both the believer and the agnostic — that learning from introspective reports means relying on a private method of observation. The idea of a first-person science misconstrues the only way we have — either in commonsense or in science — to learn from introspective reports, which is by observing public reports and subjecting them to public scrutiny. When learning from introspective reports is properly construed, introspective reports are no longer sources of unverifiable first-person data. They are public sources of public data, which are no less problematic than any other public data. Hence, using introspective reports does not fall outside of ordinary third-person science.

In conclusion, scientists need not be either blanket agnostics or blanket believers in introspective reports. They should rather be *connoisseurs* — experts who can finely judge, on public grounds, which kinds of introspective reports under which circumstances are informative about which mental states. A connoisseur is better trained and skilled, and therefore forms more accurate and reliable conclusions, than an amateur. But a connoisseur's knowledge is gained by the same public methods that inform the amateur. A connoisseur of introspective reports should do explicitly and rigorously what we attempt to do implicitly when we learn from each other's reports in our everyday life.

II: Mind Reading

Introspective reports are informative. We ask each other how we feel, what we're thinking about, how things look to us, and we learn from our answers. Our ability to learn from introspective reports is part of our *mind reading* — the ability to discover the content of other minds. This section focusses briefly on some aspects of mind reading, to see how learning from our neighbours' introspective reports might be justified.

First, we can learn about the content of minds from nonlinguistic perceptual input. Almost from birth, we respond to smiling adults whose eyes are pointed in our direction by smiling back. After a few months, we understand — from people's behaviour — what others are attending to and what their goals are. For instance, when mother turns her head suddenly to her right, her baby turns in the same direction and tries to locate whatever she must have seen. (But if a box turns to the right, a baby stares at the box and looks puzzled by its funny behaviour.) In a few years, we develop a sophisticated understanding of people's perceptions, desires, beliefs and other mental states. There is evidence that many of these skills are pre-linguistic: for instance, we share many of them with other animals. In short, we have a natural ability to respond to the content of other minds from nonlinguistic perceptual input.[2]

As members of a linguistic community, we have a mentalistic vocabulary in common. We know what it means to *feel*, *perceive* and *think*; to *believe*, *hope* and *fear*; to sense *pain*, *itch* and *pleasure*; to be *focussed*, *distracted* and *bored*; to feel *gloomy*, *excited* and *enamoured*; etc. Any competent speaker of our language knows how to apply mentalistic predicates, what mentalistic predicates name what conditions, and what inferences can be drawn from statements that contain mentalistic predicates. Any individual who shares those abilities with us, we say, *understands* our mentalistic language.[3] Those who understand our mentalistic language can share their information about minds, theirs and others', through first-person and third-person reports. Embodied within our linguistic competence, we inherit a lot of useful information that we could never acquire by dealing with nonlinguistic inputs: wisdom accumulated by our ancestors through millennia of dealing with minds. Furthermore, every mentalistic term, say *jealousy*, embodies within its meaning psychological assumptions, in this case about people's propensity to jealousy under appropriate circumstances. These assumptions are a precious part of the commonsense mentalistic beliefs of anyone who understands mentalistic terms.

With perception and language in place, and thanks to our propensity to form beliefs, we form personal beliefs about other minds, and we share psychological information with them. We observe our neighbours' behaviour. We register what our parents, friends and teachers have to say about minds. We learn from stories about other minds, and occasionally we even read psychology books. We accumulate a large body of psychological beliefs, ranging from beliefs about our friends' personalities to beliefs about how people react to advertisements (i.e., in many cases, by forming desires). To enter the business of mentalistic statements and generalizations, to find evidence for and against them, we have no need for a

---

[2] This ability is part of what psychologists call Theory of Mind. There is a vast psychological literature, which there is no room to summarize here, devoted to the study of Theory of Mind and its scientific explanation. A good review, focussed on Theory of Mind in infancy, is Johnson (2000).

[3] Marconi (1997) offers a general theory of semantic competence along these lines.

scientific psychological theory. All we need is our perceptual skills, our linguistic competence and other cognitive capacities involved in forming beliefs.

Introspection contributes to our commonsense beliefs about minds. It helps a lot in knowing ourselves. It probably helps learn our mentalistic language and understand others in analogy with us. Perhaps it even helps develop our perceptual skills. This need not spoil the publicity of mind reading resources. Introspection informs us privately, but its behavioural output, to the extent that it contributes to our commonsense psychological knowledge, is as public as any other piece of psychological information. After we introspect, we can report our findings to the rest of our community, who trades them on the same market as the output of all of our cognitive faculties.

In the trade of introspective reports, our shared resources come especially handy. When Rebecca tells us something about her mind, we not only understand what she says but also exploit what we already believe to evaluate it. Rebecca can lie, and sometimes we spot clues that she's lying. For instance, many times we asked her what she was thinking, and her answer was, 'Nothing'. Most of those times we knew she was lying, and often we knew what she was thinking, too. Of course, we can be deceived but, in the absence of signs of lying, we take Rebecca to be sincere. This is good because, like the rest of us, she usually *is* sincere. But Rebecca can also engage in wishful thinking. Sometimes she asserts she is calm and relaxed even when she is visibly tense. In those cases, we don't take her introspective reports at face value, nor do we think she is lying to us: we correctly infer from her report, the way she looks and acts, and other evidence we have, that she is tense but delusional. Jennifer, on the other hand, lately has been in a love frenzy. She goes from boy to boy, falling in love with all. Or better: this is what she says, but we know it isn't true. We've seen her really in love once, and it wasn't like *that*. As to these recent cases, she isn't in love but she desperately *wants* to be, though she'd never admit it.

Delusion is only one of many ways in which introspective reports are defeasible. Generating accurate introspective reports is likely to require the proper interplay of a large number of cognitive processes, ranging from perceptual to inferential to motor. Any of these processes may malfunction in any number of ways, and any malfunction may affect the accuracy of introspective reports. For instance, a subject's report that something looks green to him may be faulty because his introspective faculty is damaged, or because his ability to issue reports is defective, or because he doesn't know the meaning of 'green' or of 'looks,' or because his colour perception is impaired to the point that his colour experiences of things are incommensurable to those of others.[4] We can discover whether someone is cognitively impaired on public grounds, independently of any individual introspective report. So, when our neighbours

---

[4] I have a light achromatopsia myself. Before I was diagnosed, I used to occasionally argue with people as to the colour of things: if I thought something looked green to me, I would argue that it was green. Now I defer to others, and I no longer trust *my own* introspective reports on colours: if I feel tempted to say that something looks green to me but people tell me it's blue, I conclude it doesn't really look green to me — my temptation to think so must be a side effect of my achromatopsia.

give us a report, we don't accept what they say without question — we often form an opinion about their minds that's more accurate than theirs.

Most of the time, though, we take what our neighbours are saying — more precisely, what we understand of it — to be approximately true. We take their introspective reports to inform us accurately about their mental condition. This is not because we blindly trust what they tell us but because, most times, their reports sound kosher — not fishy — to our well-trained ears. We also relate what they say to their environment and to what we know about people in general and them in particular, which usually suggests that their reports are accurate. For example, our commonsense psychological assumptions include that people *perceive* objects, that perception is necessary both for certain manipulations of objects and for generating visual reports, and that objects are perceivable only when they are in somebody's visual field. So, if Andrea says that she *sees* the salad bowl, we can (implicitly) check the reliability of her report by looking at her and the salad bowl, listening to her accurate description of the bowl's shape, colour, position, etc., and noting that she comes away from the shelves carrying the bowl we asked her to get. Under our commonsense psychological assumptions, both her report and her behaviour hardly fit with the hypothesis that she isn't seeing the bowl. Although other reports can be subtler to evaluate than visual ones, the same applies, *mutatis mutandis*, to reports of non-visual perceptions, memories, feelings, etc. Andrea's sincere introspective reports can still be faulty in some respect, for example because she is colour blind, but for the most part we have plenty of reasons to trust them. And when we don't, we distrust her reports on grounds that are public as well.

All we need for evaluating introspective reports — besides our commonsense beliefs about environments and circumstances — is the resources I briefly described above, all of which can be shared. The combination of the above, including our evaluation of introspective reports, constitutes mind reading. Like many skills, mastering mind reading takes a lifetime, and we can always improve at it. And, albeit we do it mostly automatically, it is painstaking: any new introspective report requires evaluation in its own right, given the evidence that's available in the context of its utterance.

This stands in opposition to the introspection agnostic, who thinks we should be neutral about the content of introspective reports. It also stands in opposition to the introspection believer, who thinks we must rely on the general assumption that introspection is truthful. If the believer were right, we could never establish that introspective reports are accurate or inaccurate; we should just resign ourselves to trust our neighbours' reports. But the believer's suggestion is the opposite of our experience: we start with some natural epistemic resources and, building upon them, we learn to evaluate single introspective reports — one by one. In the long run, at most we establish the accuracy of some introspective reports under some conditions and the inaccuracy of other reports under other conditions. Our mind reading skill is always improvable; our beliefs about whose reports are accurate under what conditions are always revisable. We never establish the truthfulness of most introspective reports under most conditions; we are never in a position to say: we have tested enough introspective reports; we safely agree with the believer that introspection is truthful; therefore, from now on we will trust what sincere people say about themselves without question.

We can, however, collect the information that we accumulate and ask questions or make tentative generalizations about the minds of individuals or groups: Andrea gets easily excited; does Lance tend to depression? Moya enjoys thinking about her Catholic background; Italians perceive public authority as an enemy; who's more emotional, women or men? These questions and generalizations are made, and answered or tested, by the same inferential processes that we use in other domains of our commonsense beliefs, without need for any assumption about introspection's truthfulness. These commonsense generalizations, in turn, become part of the beliefs that help us evaluate future introspective reports. In our philosophical moments, we might feel inclined to make a generalization to the effect that, *ceteris paribus*, people's introspective reports are accurate. We might even call the process that leads us from individual introspective reports to this generalization a global validation of 'introspection's truthfulness'. But this is an empty claim unless backed up by our ability to (fallibly) evaluate the accuracy of individual reports in specific circumstances.

After all I've said, we still can't introspect other minds or otherwise directly observe them, but — contrary to what both agnostics and believers suggests — we do have means to evaluate the accuracy of introspective reports. In our normal interactions, usually without being aware of it, we constantly check the validity of each other's reports by these means. This practice warrants our reliance on introspective reports to gauge other minds. For some limited purposes, such as treating patients, this may be all the warrant needed by scientists who use introspective reports as evidence. But if scientists want to use introspective reports in testing hypotheses about minds, they need a more explicit, rigorous and sophisticated version of our everyday, implicit method. In the rest of this paper, I consider how we may upgrade our use of introspective reports from commonsense to science.

III: The Epistemic Role of Introspective Reports in Science

Using introspective reports in science doesn't require treating introspective reports as a source of first-person, unverifiable data, gathered by observers practicing a private method of observation. A proper scientific treatment of introspective reports should proceed in the third person, just like the proper scientific treatment of any other sources of data.

A method is a series of operations performed on some objects and instruments — what scientists call *materials*. Any material has properties, and empirical observation methods are designed to exhibit some fact or phenomenon by exploiting some properties of some materials. Scientific observation and experimentation is largely a matter of know-how: in manipulating their materials, researchers need to know how to do what they are doing, but they need not know how their materials exhibit their properties, much less have a theory of them. The microscopist needn't know how her microscope works, nor does the anaesthetist need to know everything about how the patient is made insensitive to pain. In fact, the point of doing research is to uncover something *unknown* through one's materials. As long as materials have the properties needed for a method to work, ignorance of how those properties are generated by the materials, or even ignorance of what properties are at work, does not undermine the validity of the method.

Psychologists and neuroscientists use a somewhat special material — people. Human subjects of experiments are not in the laboratory to apply scientific methods; they are part of the materials. Researchers instruct their subjects to perform tasks, and then they record their subjects' performance. The recording can be more or less sophisticated. These days, recording rarely consists of simple perception and memory; it involves tape recording or videotaping the subjects' behavioural responses, perhaps in combination with recording the subjects' neural activity by EEG or neural imaging methods. Subjects are often instructed to give verbal reports, which may include reports about their mental states. Such reports are recorded together with any other relevant piece of the subjects' responses, and then transcribed. Records of the subjects' responses, including transcriptions of their verbal reports, are sometimes called *raw data*; they are only the first stage in the production of data properly so called — those that will appear in scientific journals (Bogen & Woodward, 1988).

Any (relevant) behavioural or neural outputs of a subject can be used to collect information about that subject's mental states. The recordings of the subjects' responses, or the transcription of their verbal reports, are analyzed according to some standard procedure aimed at extracting some relevant information. The extraction of usable data from raw data requires careful methodology, especially when many assumptions are needed. A case in point is the use of neural imaging technologies in neuroscience and psychology, which face many methodological pitfalls (Bogen, 2001). In the case of verbal reports, procedures for data extraction may be as informal as ordinary mind reading, or as formal as the automatic analysis of reports by a computer program (Ericsson & Simon, 1993). The records of these analyses are further processed, and their contents summarized in a perspicuous form, until researchers come up with pieces of information that they call *data* in their scientific reports.

There is no reason why introspective reports should be used to collect information about people's mental states any less than any other relevant human output. Just as other outputs — whether neural or behavioural, verbal or nonverbal — can be exploited as sources of information about people's mental states by judicious methodology, so can introspective reports.

In fact, it's not clear whether there is a principled distinction between introspective reports and other sources of data. First, notice that being verbal is neither necessary nor sufficient for being introspective. This is because on the one hand, verbal reports need not be about mental states; on the other hand, subjects can be instructed to give any behavioural response (e.g., pressing a button) to report on their mental states. Second, notice that being instructed to report on one's mental states is neither necessary nor sufficient for doing so. This is because on the one hand, people may fail to follow instructions to introspect for a variety of reasons (e.g., misunderstanding, malice or confabulation); on the other hand, under many conditions people volunteer information on their mental states without being asked for it.[5] 'Introspective' is a mentalistic term, which — like belief, desire and other mentalistic terms — is likely to be definable only in mentalistic terms. A scientific definition of 'introspective' will thus have to wait for a scientific theory of mind, which would include a theory of introspection. As of today, such a theory is at best a work in progress. So we don't seem to have a principled distinction between introspective reports and other behaviours. And without a principled distinction, it's unclear what a ban of introspective reports would amount to.

Fortunately, just as other scientists need not have a definitive scientific understanding of the processes on which they rely in collecting their data, we don't need to have an exact understanding of how introspective reports are generated in order to use them as sources of data. Like other scientists relying on other materials, we need not know the details of the introspective process for our use of introspective reports to be legitimate. Reliance on introspective reports to generate data should work in the same way as reliance on any other aspect of a subject's behaviour. The introspecting subjects, as part of the materials of the experiment, execute their task, which — in this case — includes giving introspective reports. The subjects' responses, reports included, are recorded. Then, records of the reports are analyzed to extract information about the mental states of subjects, just as records of other behavioural and neural responses are analyzed for the same purpose.

The method followed by scientists to generate data from introspective reports includes instructing the subjects, recording their behaviour and analyzing their reports according to appropriate assumptions and standard procedures. In the context of scientific methodology, introspection need not be a first-person 'method' for generating first-person data, but only a cognitive process whose outcome is observable behaviour, which in turn is a legitimate source of data about the subject's mental states.

A final point should be considered. Human subjects elicit introspective reports in response to instructions from human investigators, and the interaction between investigators and subjects is relevant to the form and content of introspective reports. Because of this, Jack and Roepstorff have argued that scientists eliciting and using introspective reports should adopt a 'second-person' perspective in which they treat subjects as 'responsible conscious agents capable of understanding and acting out the role intended' (Jack & Roepstorff, 2002, Box 2). I agree so far. But according to Jack and Roepstorff, this 'second-person' perspective is methodologically distinct from the third-person perspective scientists adopt towards other sources of data. I disagree here. Human subjects are different from other experimental materials, and as such they require specialized

---

[5] In fact, according to some researchers, even certain non-linguistic creatures, such as some primates, can be made to exhibit behaviours that deserve to be construed as introspective reports.

knowledge and attitude on the part of the experimenters. Dubbing this specialized knowledge and attitude 'second-person' may be a good way to acknowledge that our fellow humans are capable of responding to us in extraordinarily subtle ways. None of this requires that experimenters step outside third-person science. On the one hand, Jack and Roepstorff's considerations apply not only to introspective reports but also to many other uses of human subjects in empirical research. On the other hand, extracting data from *any* materials — not only from humans — must be based on specialized knowledge of the materials. So taking a 'second-person' perspective towards experimental subjects is not in contrast with taking a third-person perspective towards the subjects' reports and the data extracted from them. The remaining question is how to validate the data.

IV: How to Validate Data from Introspective Reports

When scientists collect data, they face at least three distinct issues of validity.[6] Collecting data from introspective reports results in the same issues, which can be dealt with by the same strategies that scientists rely on in collecting any data.

First, raw data must correlate with a genuine phenomenon and not be an experimental artefact. This issue may be called *process reliability*. If the instruments are imprecise, if some variable is not controlled for, if some unpredicted condition defeats the intended outcome, then the raw data — no matter how carefully handled — will mislead the investigators. For example, if a cell line is contaminated by the wrong kind of cells, it will yield useless data. Scientific chronicles are replete with anecdotes of bogus effects generated by unexpected confounding factors. A lot of scientific ingenuity goes into minimizing these risks — it's a never-ending struggle. So, when a process or tool, like a microscope or telescope, becomes standard in a discipline, usually scientists study its properties systematically to determine under what conditions it can be effectively exploited in research. For example, if a researcher wants to anesthetize animals, she needs to know, for any given drug, animal and experiment, what dosage to administer, whether the experiment will interfere with anaesthesia, and whether anaesthesia will interfere with the experiment. That information is specific to anaesthesia, and useless when anaesthesia is not part of an experimental method. Creating a correlation between a measurement outcome and a genuine phenomenon requires a lot of specialized knowledge and skill — there is no general recipe.

Solving the problem of process reliability is mainly about preparing the materials in a careful way, learning about confounding factors (often by trial and error) and taking precautions against them. In the case of introspection, this involves formulating effective and unambiguous instructions for the experimental subjects, recognizing specific confounding factors — such as lack of cooperation, delusion, confabulation, etc. — and learning to avoid those confounding factors. Dealing with the specific confounding factors that affect introspective

---

[6] Validity should not be confused with verifiability. Statements, such as introspective reports, may be verifiable or unverifiable; data, such as data extracted from introspective reports, may be valid or invalid.

reports requires specialized knowledge and a systematic study of introspection, and it may be quite difficult to make all the relevant knowledge explicit. But there is no principled difficulty.

A further upgrade towards a more scientific use of introspective reports would be the identification of mechanisms and processes responsible for generating introspective reports and of ways in which these mechanisms can malfunction. Neuroimaging studies, as well as lesion studies and other traditional techniques, are likely to help in this respect.[7] In principle, knowing the neural correlates of introspection can help determine whether a behaviour is or isn't a genuine introspective report. If and when we have robust results about what mechanisms and processes are responsible for introspection, we could legitimately conclude that when the wrong mechanisms or processes are involved, the ensuing reports are not introspective.

A second validity issue pertains to mining the raw data for information. This issue may be called *extraction publicity*. Even if the raw data do correlate with a genuine phenomenon, processing them might add unwarranted or question-begging assumptions that skew the results. A classical example is the case of René Blondot and other physicists of his time, whose apparatus was designed to reveal the presence of a new type of radiation called N-rays. Since N-rays do not exist, the raw data generated by the apparatus correctly correlated with the absence of N-rays. Some physicists who observed raw data from similar apparata registered no effects of N-rays. But Blondot and some colleagues, who were convinced N-rays existed, claimed they were seeing the effects of N-rays — and proceeded to publish dozens of papers about them. Their illusion was publicly exposed by physicist R.W. Wood, who visited Blondot's laboratory in 1904. While Blondot was putatively measuring the effects of N-rays, Wood surreptitiously removed a crucial part of Blondot's apparatus, without Blondot noticing any diminution in the intensity of the N-rays' effects (Klotz, 1980; Nye, 1980). To reduce the risk of spoiling data during their processing, data must be analyzed and processed according to standard procedures that others can reproduce.

In psychology, to insure extraction publicity in the face of the qualitative nature of many raw data, the data analysis is often done by two different observers studying experimental records independently of each other, without knowing what hypothesis is being tested by the principal investigators. If the two observers obtain the same results from the same raw data, the information they extract is considered trustworthy.

To make the extraction of data from introspective reports as public and reproducible as possible, similar precautions should be taken. In their methodological study of introspective reports, Ericsson and Simon mention some of the important ones (Ericsson & Simon, 1993, pp. 4–5, 276–8, 286–9): (1) researchers should record the reports rather than take selective notes; (2) recordings should be transcribed verbatim; (3) theoretical assumptions used while analyzing the reports should be as few and as weak as possible; (4) all assumptions should be

---

[7] There is a growing literature that appears to be relevant here. For example, see Frith *et al.* (1999) and Vogeley & Fink (2003).

explicit; (5) categories that are used to analyze reports should be chosen *a priori*, before the analysis takes place, on the basis of a formal analysis of the task; (6) each report should be analyzed using only information contained in that report, independently of the surrounding reports, (7) the method of analysis should be constant from one report to another; (8) individuals analyzing the reports should be blind to the hypotheses being tested; (9) different individuals should analyze the same reports independently, and their outcome should be the same.

All these procedures are analogous to those used for other kinds of behavioural data, like eye movements and fixations. It may not always be possible or even appropriate to follow all of these precautions. In searching for theories in new domains, for example, the data analysis may be less formal than this methodology suggests, but whenever possible — especially when testing theories — the more rigorous method should be followed (cf. Ericsson & Simon, 1993, p. 6).

A third validity issue pertains to what the data correlate *with*. The phenomenon being measured and investigated must be properly understood and described. This issue, which may be called *framework validity*, is part and parcel of the science that is being created during the investigation. Different researchers — working under different theoretical assumptions — may genuinely disagree about what a method is measuring. For instance, when particle physicists announce the results of their data analyses, there may be a period of genuine controversy over what they have measured; for example, as to whether a new particle has been discovered, what particle it is, and what its properties are. These controversies are normally resolved by a combination of further experimentation and theorizing. The former yields more data about the phenomenon, while the latter aims at either accommodating the new findings within the existing conceptual frameworks or extending existing frameworks so that they can accommodate the new findings. In the end, hopefully an uncontroversial interpretation of the data emerges.

To determine what introspective reports correlate with is not straightforward. Answers that have been offered in the literature include the contents of short-term memory during a task execution, the contents of long-term memory traces of past mental states of the subject, the subject's representation of herself, the subject's qualitative state of consciousness, and what the subject believes the experimenter wants to hear. (These answers are not always mutually exclusive.) Different kinds of introspective reports may correlate with different phenomena under different conditions. For example, reports issued during the execution of a task may correlate with the contents of short-term memory, whereas reports issued after a task is completed may correlate with the contents of long-term memory. Different experimental paradigms may be able to exploit those correlations for some purpose or another. The answer to these questions is for empirical science to find. It will require both a theoretical language in which to couch the description of what is being measured and a lot of empirical investigation.

A model for how to proceed in the matter of framework validity is the classic study by Ericsson and Simon (1993; see also Ericsson, 2003). Ericsson and Simon developed a rigorous methodology for extracting data from two kinds of introspective reports pertaining to the execution of cognitive tasks. Reports of the first kind — *concurrent* introspective reports — are issued during the execution of a task. Reports of the second kind — *retrospective* introspective reports — are issued after a task has been executed.

In determining what introspective reports correlate with, Ericsson and Simon adopted the following framework, which they saw as supported by decades of cognitive psychology. Human cognition is information processing. Information structures, often coming from perceptual input, are acquired by a central processor and temporarily stored in short-term memory. While in short-term memory, information structures are available for further processing, including the process of verbalization. From short-term memory, information can be transferred into long-term memory, from which it can be retrieved and verbalized.

Ericsson and Simon used their framework to determine what introspective reports correlate with. According to them, verbalization encodes information structures into concurrent reports — reports containing information being processed in short-term memory at the time the report is made — or retrospective reports — reports made after the time when the information they contain was processed by the subject (cf. Ericsson & Simon, 1993, p. 11).

Ericsson and Simon's theory may well, in practice, fail to account for some experimental procedures involving introspective reports. For that matter, it may well be false. Those are empirical issues of no concern here. What matters for the current argument is that their theory is supported by evidence collected by public methods, which can be applied by their colleagues. Other researchers may disagree, but not accuse Ericsson and Simon's theory of being based on untestable assumptions, e.g. that introspection is truthful.

Ericsson and Simon do not stop at formulating a rigorous methodology for extracting data from introspective reports. They actually offer evidence that the kinds of reports they study offer reliable information about the mental states of subjects. They do so in two steps. First, they exploit their framework to predict what introspective reports should be like under certain circumstances; second, they review three sources of evidence showing that their predictions are correct. The first source of evidence is independent of introspection itself, while the other two compare the information contained in different kinds of introspective reports (Ericsson & Simon, 1993, p. 137).

First source. Many cognitive tasks require the use of perceptually available information. For example, if a subject performing arithmetic calculations with paper and pencil reports that she is summing up two numbers while her gaze is shifting between two digits on paper that represent those numbers, the eye fixations of the subject — combined with the result of her calculation — are evidence of the accuracy of the subject's report. This evidence can be gathered because the same information that is perceived by the subject is also available to the experimenters, and can be recorded by a camera together with the subject's eye fixations. This is far from trivial: in the case of Anton's syndrome — unawareness of one's own blindness — subjects describe what they claim to be seeing while their eyes point in a certain direction, but there's no correspondence between the content of their reports and the stimuli in their visual field. In normal cases, though, there is an experimentally established correlation between eye fixations and the information contained in introspective reports. So, via the theoretical assumption that eye fixations tell what information is processed in short-term memory, eye fixations give researchers public means — independent of introspection itself — to establish the accuracy of introspective reports (Ericsson & Simon, 1993, p. 173).

Second source. Ericsson and Simon's theory predicts that, if identical cognitive processes are reproduced, introspective reports made during those processes will contain identical information. Subjects can perform a task and give reports on it more than once. If each time their reports are the same in relevant respects, this identity of reports supports Ericsson and Simon's theory, and in turn the conclusion that the reports are accurate. It may be difficult to reproduce exactly the same cognitive process in the same subject, because learning and memory intervene between each two executions of a task, making slight changes to the cognitive process. But some tasks, like mental arithmetic, do reproduce the same processes each time they are executed, and the results corroborate the reliability of introspective reports (Ericsson & Simon, 1993, pp. 356–7).

Third source. Introspective reports can be either concurrent or retrospective relative to task execution. Concurrent reports draw on short-term memory, whereas retrospective ones draw on the trace of a process that's stored in long-term memory. By asking subjects to give both concurrent and retrospective reports on the same task, researchers can compare the content of two sets of reports pertaining to the same cognitive process. Ericsson and Simon's theory predicts that retrospective and concurrent reports will be mutually consistent, and that the former will contain less information than the latter. The authors review experimental evidence showing that their predictions are fulfilled (Ericsson & Simon, 1993, p. 357ff).

Ericsson and Simon's rigorous theory of introspection, and their consequent validation of concurrent and retrospective reports, is a valuable contribution to the methodology of introspective reports. Other scientists using introspective reports to gather data need not subscribe to Ericsson and Simon's theory, nor do they need to have an alternative theory of introspection. As long as they have no reasons to believe their raw data are experimental artefacts, and as long as they are following public procedures of data extraction, they should feel free to (fallibly) search for the best interpretation of their data.

V: Conclusion

Our ordinary mind-reading skill allows us to understand, evaluate and interpret introspective reports so as to extract useful information about our neighbours' minds. This commonsense treatment of introspective reports is wiser than either agnosticism or faith. It shows that we can learn from introspective reports without needing to construe them as the output of a private method of *observation*. Introspection can be construed as a cognitive process yielding a kind of behaviour, which can yield information like any other process and behaviour. When this is done, the supposed 'unverifiability' of introspective reports becomes a non-issue. If we establish what information can be extracted from which sorts of introspective reports under what circumstances, there is no reason why we shouldn't use them as sources of scientific data.

For some scientific purposes, our ordinary mind reading, perhaps supplemented by the results of relevant scientific research, might suffice to underwrite the extraction of information from introspective reports. For instance, neuropsychologists routinely resort to this practice in interpreting their patients' reports. But relying on our ordinary mind reading is not enough for many scientific purposes, especially the rigorous testing of hypotheses in psychology and neuroscience. A proper scientific methodology of introspective reports should go beyond ordinary mind reading. Scientists can take our commonsense treatment of introspective reports and make it more explicit and rigorous.

Scientists using introspective reports face the same problems that other scientists face, and they can solve them by specialized versions of the strategies that other scientists employ. They need to make sure their raw data are not experimental artefacts, by learning about factors that confound their experimental results and avoiding them. They should steer clear of question-begging assumptions while processing their data, by following public procedures that others can replicate. And they should discover, by a mixture of empirical research and theoretical construction, what mental states correlate with what introspective reports under what conditions.

None of the above constitutes a special 'first-person' science. The proper scientific use of introspective reports falls well within the boundaries of ordinary third-person science. This is what makes the use of introspective reports scientifically legitimate, and why scientists should have no blind faith in introspective reports.

Introspection's truthfulness is not something to be established by armchair philosophizing; it takes hard empirical research to find out what can be learned by what reports under what circumstances. We can discover what to infer from introspective reports in a similar way to how we do it every day, but with the extra knowledge, methodological care and precision that comes with scientific study. In doing so, we should rely on our best-established scientific theories of mind and brain and our specific knowledge of the circumstances in which the reports are generated (the environment, the task being performed and the instructions given to the subject). We should also develop empirical tests of the reliability of introspective reports. A serious start in this direction has already been made, at least for certain kinds of reports (Ericsson & Simon, 1993; Ericsson, 2003). Sorting out the use of introspective reports as sources of data is going to be a painstaking, piecemeal task, but it promises to enhance our science of the mind and brain.

The generic term 'introspective reports' blurs important distinctions between different kinds of reports. There are concurrent reports, which are generated while performing a task, and retrospective reports, which are generated from memory traces. There are reports that can be issued automatically and reports that require attention. There are reports about tasks whose execution can be independently observed and reports about qualitative states, like pain. Not all introspective reports are informative in the same way, and not all of them present the same methodological difficulties. A proper methodology of introspective reports will have to take all these differences into account.

Acknowledgements

I thank Jim Bogen, Matt Boyle, Carl Craver, Peter Machamer, Andreas Roepstorff, Andrea Scarantino, Becka Skloot and two referees for helpful comments and discussions on this topic. Special thanks to Tony Jack for his specially insightful comments.

References

Bogen, J. (2001), 'Functional imaging: Some epistemic hot spots', in *Theory and Method in the Neurosciences*, ed. P. Machamer, R. Grush and P. McLaughlin (Pittsburgh: University of Pittsburgh Press), pp. 173–99.

Bogen, J., Woodward, J. (1988), 'Saving the phenomena', *Philosophical Review*, **XCVII** (3), pp. 303–52.

Chalmers, D.J. (1999), 'First-person methods in the science of consciousness', *Consciousness Bulletin*, **Fall**, 1999.

Chalmers, D.J. (1996), *The Conscious Mind: In Search of a Fundamental Theory* (Oxford: Oxford University Press).

Dennett, D.C. (1991), *Consciousness Explained* (Boston, MA: Little, Brown & Co.).

Dennett, D.C. (2001), 'Are we explaining consciousness yet?' *Cognition*, 79, pp. 221–37.

Ericsson, K.A. (2003), 'How to elicit verbal reports that provide valid unobtrusive externalization of concurrent thinking?', *Journal of Consciousness Studies*, **10** (9–10) pp. 1–18.

Ericsson, K.A., Simon, H.A. (1993), *Protocol Analysis* (Cambridge, MA: MIT Press).

Frith, C., Perry, R., *et al.* (1999), 'The neural correlates of conscious experience: An experimental framework', *Trends in Cognitive Sciences*, **3** (3), pp. 105–14.

Goldman, A.I. (1997), 'Science, publicity, and consciousness', *Philosophy of Science*, **64**, pp. 525–45.

Goldman, A.I. (2000), 'Can science know when you're conscious?' *Journal of Consciousness Studies*, **7** (5), pp. 3–22.

Jack, A.I., Roepstorff, A. (2002), 'Introspection and cognitive brain mapping: From stimulus–response to script–report', *Trends in Cognitive Sciences*, **6** (8), pp. 333–9.

Johnson, S. (2000), 'The recognition of mentalistic agents in infancy', *Trends in Cognitive Sciences*, **4**, pp. 22–8.

Klotz, I.M. (1980), 'The N-ray affair', *Scientific American*, **242** (5), pp. 168–75.

Lyons, W.E. (1986), *The Disappearance of Introspection* (Cambridge, MA: MIT Press).

Marconi, D. (1997), *Lexical Competence* (Cambridge, MA: MIT Press).

Nye, M.J. (1980), 'N-rays: An episode in the history and psychology of science', *Historical Studies in the Physical and Biological Sciences*, **11** (1), pp. 125–56.

Piccinini, G. (forthcoming), 'Epistemic divergence and the publicity of scientific methods', *Studies in the History and Philosophy of Science*.

Price, D., Aydede, M. (submitted), 'The experimental use of introspection in the scientific study of pain and its integration with third-person methodologies: The experiential–phenomenological approach'.

Vogeley, K., Fink, G.R. (2003), 'Neural correlates of the first-person perspective', *Trends in Cognitive Sciences*, **7** (1), pp. 38–42.

Richard E. Cytowic

The Clinician's Paradox
Believing Those You Must Not Trust

Clinicians have a convention whereby **symptoms** *are subjective statements 'as told by' patients, whereas* **signs** *are outwardly observable facts. Yet both first-person reports and third-person observations are theory laden and can bias conclusions. Two aspects of the oft-mentioned unreliability of reports are the subject's interpretation of them and the experimenter's assumptions when translating introspective reports into scientifically useful characterizations. Meticulous training of introspectors can address their mischief, whereas investigators can become more attentive to their own theory-laden biases. In the case of hallucinations, for example, ignoring some customary third-person constructs and focusing on the visual experience itself has led to fresh explanations of visual symptoms based on cortical physiology rather than conceptual categories. Constructs that historically have ignored the subject's state of mind are also problematic; an example is the so-called resting state during metabolic brain imaging, long believed to reflect a blank mental slate.*

Introspective reports, not accepted literally but properly interpreted and revised by investigators as necessary, are legitimate sources of data.

When practising as either a neurologist or neuropsychologist, a great deal of data upon which I draw inferences and make decisions is clinical, understood as meaning generally subjective and hence unreliable in the eyes of an ideal objectivist. Clinical convention conceptualizes *symptoms* — such as pain, vertigo or forgetfulness — as subjective statements 'as told by' patients, whereas *signs* — such as paralysis, seizure or agrammatical errors — are objective, outwardly observable facts.

In practice, the clinician's milieu is muddled. Not much of what we have to work with is purely objective because we are often not around to witness matters first hand and so must reconstruct events, sometimes from second-hand sources. Looking at the other extreme, we realize that even the most subjective symptoms, such as pain, are coupled to objective signs (in this case, say, increased sympathetic nervous activity). Neurologists and kindred clinicians, like me, wonder if the subjective–objective dichotomy that so exercises philosophers is overwrought. The distinction does not appear sharp to us because we operate within a paradox: we must rely on patient reports, yet cannot trust them fully.

Part of the problem is that patients frequently *interpret* events instead of *reporting* them straightforwardly as one would wish ideally. As this is an occupational fact, clinicians have traditionally learned to look beyond face value when translating first-person accounts into a third-person understanding of how the nervous system operates. In so doing, however, examiners are prone to bias introspective reports with their own theory-laden assumptions.

Consider an anecdote and two responses to it, which I caricature as the objectively oriented American and the subjectively oriented British. The anecdote is that a well-known newspaper editor falls at home and now seems 'different' to his worried family. The editor protests that he is fine and wants to go home. The Yank, whose schedule runs late as usual, pops into the exam cubicle, hears the wife's statement, 'he's confused', and responds, 'let's get a scan', dispatching the patient into the tunnel of technology without further ado.

Our Brit is schooled in the 'clinical method' as exemplified by the clinicopathological approach taught at a renowned institution such as Queen Square. The editor's reputation and implied level of education predisposes the Brit favourably. He judges the patient's comportment and conversation to be normal; in fact, his vocabulary sounds quite rich. What could the wife mean? Looking for evidence of confusion but finding none, the Brit's empirical knowledge that vocabulary does not falter until late in the course of many brain pathologies, when physical signs may be already apparent, makes him discount the smooth talk and probe further. Asked the reason for falling, the editor says, 'I lost my glasses . . . I was going to the garage'. Incongruent remarks are exactly the kind of signal our Brit has learned that signify something amiss in cognitive or language domains (e.g. a dementia or aphasia). When drawn out of him, the editor's narrative turns implausible, fragmented, and shows that he does not grasp the meaning of the immediate situation. A social history from the wife uncovers heavy drinking and skipped meals. Next, the suspicion of anterograde and retrograde amnesia is quickly confirmed by a simple procedure, as is the presence of ataxia when tandem walking is tested — our clinician is now confident that ataxia caused the fall and that the mental disorder has gone unnoticed until now, a textbook example of 'decompensation'.

Unlike his American colleague who asked for an objective test straightaway, the British clinician applies his knowledge base and examination manoeuvres flexibly as the clinical situation unfolds, and so uncovers Wernicke-Korsakoff syndrome in the editor. His prepared mind was not mislead by statements, but deduced a diagnosis (Latin: *dia* = through + *gnosis* = knowledge) that the editor is a so-called maintenance alcoholic with thiamine deficiency causing brain damage that conforms to a recognized syndrome. Should this clinician request a scan, he will be alert for specific lesions in the diencephalon or note that cerebral atrophy exists in a *recognized pattern* involving the vermis, interhemispheric fissure and temporal fissures (Victor *et al.*, 2000).

When the Yank's scan report comes back saying, 'mild atrophy', he may set off on an expensive fishing expedition of more tests to follow up this objective but unhelpful finding, whereas the Brit may be satisfied with his diagnosis and search no further, although possibly at the cost of failing to uncover a different etiology for the editor's condition. This possibility is small, although time will tell. Note also that the disinterested radiologist, assumed to be objective in the sense of being a neutral third party, is like the Yank in having no context 'in what way' the patient is 'confused', as the requisition slip states. His anatomical interpretation of the scan is limited to a general statement. Provided context, however, he will *view the image differently* now that he knows what to look for, perhaps saying that the pattern of atrophy observed is 'consistent with alcoholic cerebral degeneration', a clinically more useful statement to all concerned.

These caricatured scenarios surrounding a common syndrome raise issues of preparation, expectation, assumption and social context. Such factors can influence both subjective and objective observations. Examples supporting the absolute fallibility of verbal reports are easy to come by. For example, split-brain individuals demonstrate the existence of cognition not accessible to language as well as serving as an instance wherein verbal reports can actively mislead both patient and clinician (Zaidel *et al.*, 2003). Counterexamples claiming verbal reports as crucial to scientific understanding are also plentiful. Dreaming during REM sleep is an obvious one: if no one had awakened sleepers during this objective yet enigmatic EEG event, the meaning of REM sleep would have remained undeciphered.

Although my scenarios above refer to the intellectual process by which clinicians make medical judgments (an area with its own literature), I will not focus on it. Rather, I was invited to comment on *the necessity of clinicians and experimenters to adopt a different interpretation from that of their subject*. This implies that clinical reports can never be accepted verbatim; rather, their meaning must be drawn out and elicited. To use Henry James's term, they are 'rendered', involving the clinician as much as a participating artist as he is a detached amanuensis. After all, there must be some reason behind the phrase 'art of medicine'. Does it help, or get in the way?

I consider that (1) introspections are subject to two kinds of biases: (a) subjects interpret their experiences instead of factually reporting them, and (b) investigators fail to recognize that their assumptions and interpretations are theory-laden. As a corollary, scientists cannot accept introspective reports literally. Accordingly: (2) (a) can be addressed by training subjects; (3) both (a) and (b) can be lessened by using a script; and (4) other sources of so-called objective data (e.g. neural metabolism) are also biased by (b), with no reason to suppose that (b) is any more problematic for introspective reports than it is for other kinds of data.

It follows that introspective reports, not accepted literally but properly interpreted and revised by investigators as necessary, are legitimate sources of data.

Jack and Shallice (2001) comment: 'Introspective evidence always arrives interpreted... all descriptions of experience, no matter how basic, carry implicit theoretical commitments of one sort or another.' They concur with the general statement, typically said with regard to objectivity, that observation in support of theories is itself always theory-laden. There are ways to address this when assessing first-person reports, particularly those involving perceptual issues where the facts of physiology are often counterintuitive to common sense (so-called 'folk psychology' accounts of vision, for instance, are ignorant of the vertical meridian or the quadrants in the visual fields, indeed of the very concept of a visual field).

Consider the experience of visual hallucinations. In describing Heinrich Klüver's work starting in the 1920s and spanning many decades, I recounted how he was initially 'frustrated by the vagueness with which subjects described their experience, their eagerness to yield uncritically to cosmic or religious interpretations, to 'interpret' or poetically embroider the experience in lieu of straightforward but concrete description, and their tendency to be overwhelmed and awed by the 'indescribableness' of their visions' (Cytowic, 1997, p. 29). He intuited that the novelty of what subjects saw and the vivid colours grabbed their attention more than did the spatial configurations. Because people ordinarily have no need to describe their conscious experience, they do not pay attention to its subtle features. By training his observers, however, Klüver (1942; 1966) eventually did succeed in identifying three categories of visual pathology. Among these are his well-known 'form constants' encompassing four varieties of consistent hallucinogenic configuration: gratings and honeycombs, cobwebs, tunnels and cones, and spirals. (His other two categories regard variation in the number, size and shape of objects, and variation in their spatial and temporal relations.)

Others replicated and extended Klüver's work (Siegel and Jarvik, 1975; Horowitz, 1964; 1975). Recently, ffytche and Howard (1999) confirmed these form constants and other visual experiences across a range of clinical conditions, showing how 'similarities between the visual experiences of unrelated clinical and experimental contexts' are meaningful in neuroscientific terms. As Klüver foreshadowed, 'the occurrence of these [identical] symptoms in aetiologically different conditions suggests that we are dealing with some fundamental mechanisms involving various levels of the nervous system'. To achieve this, ffytche and Howard ignored customary third-person constructs of perception that are typically used to describe visual pathology — these being examples of investigator bias (b). For instance, they did not distinguish between hallucinations (experiences that have no afferent input) and illusions (experiences arising from falsely interpreted afferents), or between hallucinations wherein insight is preserved (i.e. patients know that they are hallucinating, as in the Charles Bonnet syndrome) and those where insight is lost (e.g. in schizophrenia, where patients cannot grasp the unreality of their visual experience). By ignoring these traditional theoretical constructs of how patients are 'supposed to' experience their hallucinations and focusing instead on the final common pathway of the visual experience itself, the authors were *able to identify three new categories of abnormal vision*. Noting next that they observed identical abnormalities in etiologically different conditions and experimental settings, they related each new category of visual experience to known visual cortex physiology and were able to offer a biologically based classification of positive and negative visual symptoms — something the customary third-person constructs had failed to do.

These examples stress two aspects of the oft-mentioned unreliability of reports. The more familiar is the subject's bias (a): I use the term 'embroider' to describe how they explain, interpret or rationalize their experience based on assumptions usually very different from those of neuroscience. The other side concerns the experimenter's assumptions (b), either individually or as a member of the collective neuroscience community, when translating introspective reports into scientifically useful characterizations.

The opportunity to embroider is especially ripe in synaesthesia because the experience is so ineffable, hence difficult to convey. Subjects often resort to metaphor when describing 'What it is like'. For example, my index case MW, in whom taste and smell evoked tactile perceptions predominantly in the hands and face, described the taste of spearmint as 'cool glass columns'. Was he being metaphoric, or verbally interpreting a sensory experience by translating tactile sensations into images I could understand? I sought to tease this out by asking him to describe the exact sensations he felt, rather than 'what it is like'. When given the stimulus, MW rubbed his fingertips together and moved his hand through the air as if palpating an actual object, saying:

> I feel a round shape. There's a curvature behind which I can reach, and it's very, very smooth. So it must be made of marble or glass, because what I'm feeling is this incredible satiny smoothness. There are no ripples, no little surface indentations, so it must be glass because if it were marble, I would be able to feel the roughness of the stone or the pits in the surface. It's also very cool so it has to be some sort of glass or stone material because of the temperature. What is so wonderful is the absolute smoothness of it. I can run my hand up and down, but I can't feel where the top ends. I feel that it must go on up forever. So the only thing I can explain this feeling as is that it's like a tall, smooth column made of glass . . . There is this funny sort of feeling of being able to reach my hand into this area. It's very, very pleasant. (Cytowic, 2002, p. 33).

MW experienced simple tactile qualities that he sensed as identical over time if given the same stimulus. Querying other synaesthetes about their experience together with test-retest situations subsequently confirmed this stability, thus leading to a general statement that became one of my five diagnostic criteria for the trait, namely that 'synaesthetic percepts are generic and consistent', meaning that what is experienced is never complex or pictorial, but elementary — blobs, lattices, cold, rough, sour, zigzag or geometrically simple — and essentially invariable (Cytowic, 2002, pp. 67–9). Teasing out this phenomenal feature therefore contributed to synaesthesia's nosology (the classification of diseases), and as an axiom it has held up over time.

Getting past (a)-type biases to clarify 'what it is like' led me to concurrently rethink the mechanism of synaesthesia from a wholesale transgression of sense modalities into one another's territories (as it had always been described) to the coupling of very specific attributes (what I originally called 'sensory fragments' but later recognized as the philosopher's 'qualia') in anomalous combinations: the whoosh of a furnace ignition, for example, always being *heard* as 'a stack of pink, green, and steely blue lines' about a foot in front of the subject 'moving up to the right as they fade' over the span of three seconds. Note the specificity of the description and the multiple sensations within it.

The issue of qualia led me to the binding problem (by one definition, synaesthesia is anomalous binding) and similar contradictions that synaesthesia presented to established concepts, among them modularity and functionalism. In fact, the history of synaesthesia has been an increasingly unavoidable contradiction of traditional ideas about how the nervous system is organized. The typical response of contemporary scientists was denial of synaesthesia as a real perceptual phenomena — an extreme example of (b)-type bias. Cytowic (2002a) summarizes how the inadequacy of traditional hierarchical models of brain organization led me to alternative models, such as selectively distributed systems as well as how the phenomenon may help elucidate the neural basis of metaphor, whereas Gray *et al.* (2002) discuss synaesthesia's challenge to functionalism.

Had I accepted MW's descriptions at face value or as metaphoric, rather than probing to tease out the sensory qualities he was struggling to convey, synaesthesia might have remained relegated to a mere curiosity rather than being recognized today as an anomaly with important implications for concepts of how the brain is organized (Cytowic, 2002; Smilek and Dixon, 2002; Ramachandran and Hubbard, 2001). Rather than second guessing experiential reports, the focus of those who are currently exploring the phenomenon has been on trying to understand synaesthetes' reports and seek behavioural correlates of the subjective claims. This requires interplay between first person and third person accounts. A dialogue or structured questioning between clinician and subject constitutes a second person relation between shared knowledge about experience. This kind of feedback sometimes leads to further self-observations from a synaesthete and the subsequent deployment of additional third person tests.

Theoretical constructs ((b)-type biases), though evidently derived from third-person observation, nonetheless have historically shortchanged first person accounts. My own awareness of the subject–examiner mismatch reaches back to ophthalmologic training in the 1970s. I was struck by how often patients spoke of 'seeing things' that they could describe in detail sufficiently nuanced so as to sound similar to other patients' 'things', yet the neuro-ophthalmologic exam revealed nothing amiss despite our using plenty of equipment. Faculty and trainees assumed that every optical symptom had a physical correlate that we could observe through our optics. This conceit never had to be taught explicitly because it was just taken for granted. Looking back, that confidence is breathtaking, but many (b)-type assumptions appear so 'obvious' that their bias is never questioned. Patients did not want to hear, 'I don't see anything wrong with your eyes', because it left their questions unanswered, and their persistent asking left us frustrated. We each had different expectations. Perhaps my sensitivity to the mismatch between subject–observer expectations left me open to MW's experience instead of dismissing his synaesthesia as impossible, as my peers automatically had (Cytowic, 2002, p. 64). (b)-type bias can therefore influence the very kind of data an investigator will take up as interesting.

Can we dismiss subjects' interpretations without dismissing either them or their experience, thereby leaving us empty handed of potentially valuable data? Yes, nondismissive disregard is a learnable skill so long as we acknowledge that knowledge has limits and we see conventional wisdom as porous to new ideas rather than fixed and impervious. It is another bias not to realize that human nature makes us emotionally attached to our ideas (as in the above denial of synaesthesia being real because traditional models could not account for it). Training ourselves is the compliment to training introspectors. By letting experiential reports guide data analysis, for example, the pathophysiological loci of the above-mentioned 'spontaneous visual phenomena' (as those visual 'things' are now called) was eventually determined (Lepore, 1990), just as ffytche and Howard discovered new explanations for visual hallucinations by this same approach a decade later (see above).

Reporting on my experience with synaesthesia, I cautioned: 'Though synesthetes are often dismissed as being poetic, it is *we* who must be cautious about unjustifiably interpreting their comments' (Cytowic, 1997, p. 24). Whereas training introspectors is one way to address their bias (Varella, 1996; Varella and Shear, 1999), examiners still need to become more attentive to their theory-laden assumptions that influence diverse sources of data. One theoretical construct that illustrates how an assumption can gradually change is the 'resting state' observed during metabolic brain imaging.

For twenty years it was accepted fact that subjects' brains were essentially blank when not performing the experimental stimulus task. That is, the metabolic landscapes obtained while quietly lying in the apparatus were taken as 'zero'. These resting states were even regarded as comparable between subjects. In the earliest days of measuring regional cerebral blood flow (rCBF), techniques based on following the dilution of a tracer in the bloodstream (the standard Fick principle of physiology) required experimenters to hold a 12-gauge needle and enormous syringe in the carotid artery while repeatedly shouting to the subject, 'Relax!'. So far as I can determine from those who actually conducted such measurements, no one considered if or how arousal in a terrified subject might influence the results.

When rCBF assessment changed over to the xenon^{-133} inhalation technique in the early 1970s, it took about ten years to recognize that the elevated frontal activations routinely seen in resting landscapes were artifacts of the testing environment due to novelty and emotional arousal. Only recently have we even begun to take into account what subjects actually do as they lie in today's sophisticated scanners supposedly occupied by the experiment. We now admit that subjects grind their teeth, daydream, plan, are emotionally aroused, or engaged in a range of shifting cognitive states that have nothing to do with the experiment, that may be contrary to it, and that unquestionably affect the images obtained. Subjective mental states produce metabolic landscapes that need factoring rather than being dismissed as irrelevant to the experimental activation. We are beginning to use scripts in an attempt to isolate a given cognitive process, but this approach is not yet completely adequate.

The blank slate supposition also extended to animal work. At a World Congress of Neurology, a presenter discussing unit recording of parietal neurons in monkeys reaching for targets actually said: 'The monkey is *told to sit there* and *not do anything*'. Perhaps more remarkable is that few people caught the implicit assumption that the monkey's brain is wholly devoted to the experimental task *and nothing else*. The phrase 'behaving monkey' continues to appear in many papers, but can we even define 'behaving monkey' in a satisfactory way? As Uttal (2001) argued in his provocative book, many purported cognitive functions that are not circularly defined cannot be adequately defined at all. Leopold and Logothetis (this issue) address subjective states in monkeys.

Objectivists seem always to pine for Nagel's 'view from nowhere', that Platonic somewhere that is detached from everyone's bias (Nagel, 1986). A favourite perceptual example is the colour red supposedly being determined by viewer-independent red wavelengths of a specific measure. Though trying to separate the seer from what is seen in this manner bypasses the question of whether we can know if viewers of light of such-and-such a wavelength have the same visual experience even though they may all name it 'red' (the privacy of experience argument), this example is mistaken for a more fundamental reason. First, the basis of colour vision is widely misunderstood. It is emphatically not wavelength-based. If it were, then the colour of objects would continually change because the wavelength composition of reflected light varies with the illumination. Yet a banana looks yellow whether we view it indoors in fluorescent or incandescent light, or outdoors in bright sun or shade. *Colour constancy* is a fundamental issue in vision science, a problem that Edwin Land solved thirty years ago (Land, 1974; Cytowic, 2002, pp. 254–60, 327–36) yet an elucidation not widely known.

Colour is a property that brains assign to surfaces, not a property inherent in light and not a property inherent in objects. Area V4/V8 calculates a ratio (which remains constant as illumination varies) between wavelength composition reflected from a point and that from all surrounding points, and assigns an object a constant colour. (This means that physical mind-independent objects do not literally have dispositional properties causally corresponding to experiential states representing colour. Further, synaesthetes who experience colour in response to hearing spoken words activate their left V4/V8 (Nunn, *et al.*, 2002).) As fascinating and pregnant as colour perception is for discussion, my point is that it is always a particular someone's brain performing this ratio-taking, a brain that has a unique synaptic matrix, hence a unique knowledge base determined partly by genetically determined axonal connections and more so by experience-induced modifications in their synaptic strengths. The deep intermediate processing (i.e. everything between input and output) in human brains allows identical stimuli to trigger disparate responses depending on context, experience, consequences and so forth (Mesulam, 2000). So the privacy argument may matter in ways its proponents never realized or intended.

An attempt to eliminate the subjective role of a human observer in gathering empirical data is very much a twentieth-century phenomenon, first in physics then in psychology. In the nineteenth century, psychophysicists such as Gustav Fechner tried to formulate laws regarding sensation and perception, taking as a given that mental states exist. Even today's craze for functional brain imaging, which is supposed to be anatomically objective, starts with what one wants to verify objectively: the subject's state of mind. Because the nineteenth century Brodmann maps that people use to point to this and that spot in the brain differ in absolute coordinates from person to person, fMRI interpretations are probably unreliable when subjects are averaged in a study, yet this happens routinely. So 'red wavelength' propositions — the idea of subject-independent measurement — unfortunately do lead to (b)-type errors even in the case of our latest sophisticated technology. 'Who' gets scanned matters. Another problem with functional imaging that has received scant attention is that whereas it has impressively advanced understanding of the functional landscape in human brains, it has contributed nothing to our understanding of connectivity among various structures. We currently understand very little in the human compared to our knowledge of connectivity in other species (Mesulam, 2000). If the Brodmann maps vary from person to person, connectivity probably does as well. These issues need to be determined rather than assumed one way or the other.

The experimenter may observe, but the subject has access to experience. I have accounted only for biases (a) and (b) from subject and experimenter respectively, and my conclusion stated at the beginning — that introspective reports, not accepted literally but properly interpreted and revised by investigators as necessary, are legitimate sources of data — follows only if these are the sole sources of bias. Critics of introspection typically invoke the privacy and consequent unverifiability of reports as their main reason for rejecting them as legitimate sources of scientific evidence. It is not my intent to cover all methodological issues, but this objection strikes someone with a clinician's point of view as more academic than practical.

The clinician's milieu perhaps reminds us daily that reports are our bread and butter, sometimes the only thing we have to go on. They are not so absolutely unverifiable and private that we never have some shared point of reference. In the clinician's experience, subject and object are commingled, separating fully only in the categories that human minds invent. Concepts and models — whether of brains, cognition, or parallel universes — are by definition reductions of reality produced by human minds trying to grasp relations among too many variables. A common mistake is reifying the model and confusing it with reality. When having to deal with reality — whether in the form of sick patients or experimental subjects — the clinician by temperament is likely to forgo conceptual arguments and suggest that science is better served by striving for clear expression and understanding what both subject and experimenter 'mean by' their reports, their instructions and whatever else gets said. There is no reason to suppose that investigator bias is any more problematic for introspective reports than it is for other kinds of data, such as cerebral metabolism.

Should we trust the subject, as the editors of this volume ask? I think so, but not literally. The assumed biases of both subject and investigator can and should be turned into a useful methodology. We should listen with the mind of someone on whom nothing is lost, render the data artfully, and try not to judge too quickly.

References

Cytowic, R.E. (1997), 'Synesthesia: Phenomenology and neuropsychology', in *Synaesthesia: Classic and Contemporary Readings*, ed. S. Baron-Cohen and J.E. Harrison (London: Blackwell).

Cytowic, R.E. (2002), *Synesthesia: A Union of the Senses*, 2nd ed. (Cambridge, MA: MIT Press).

Cytowic, R.E. (2002a), 'Touching tastes, seeing smells — and shaking up brain science', *Cerebrum*, **4** (3), pp. 7–26.

Cytowic, R.E. (2003/1993), *The Man Who Tasted Shapes*, rev. ed. (Cambridge, MA: MIT Press; Exeter: Imprint Academic).

ffytche, D.H. and Howard, R.J. (1999), 'The perceptual consequences of visual loss: "positive" pathologies of vision', *Brain*, **122** (pt. 7), pp. 1247–60.

Gray, J.A., *et al.* (2002), 'Implications of synaesthesia for functionalism: theory and experiments', *Journal of Consciousness Studies*, **9** (12), pp. 5–31.

Jack, A.I. and Shallice, T. (2001), 'Introspective physicalism as an approach to the science of consciousness', *Cognition*, **79** (1–2), pp. 161–96.

Horowitz, M.J. (1964), 'The imagery of visual hallucinations', *Journal of Nervous and Mental Diseases*, **138**, pp. 513–23.

Horowitz, M.J. (1975), 'Hallucinations: an information processing approach', in *Hallucinations: Behavior, Experience and Theory*, ed. R.K. Siegel and L.J. West (New York: Wiley).

Klüver, H. (1942), 'Mechanisms of hallucinations', in *Studies in Personality*, ed. Q. McNemar and M.A. Merrill (New York: McGraw Hill).

Klüver, H. (1966), *Mescal and Mechanisms of Hallucination* (Chicago: University of Chicago Press).

Land, E. (1974), 'The retinex theory of color vision', *Proceedings of the Royal Institute of Great Britain*, **47**, pp. 23–58.

Lepore, F.E. (1990), 'Spontaneous visual phenomena with visual loss: 104 patients with lesions of retinal and neural afferent pathways', *Neurology*, **40**, pp. 444–7.

Mesulam, M.-M. (2000), 'Behavioral neuroanatomy', in *Principles of Behavioral and Cognitive Neurology*, 2nd ed., ed. M.-M. Mesulam (New York: Oxford), pp. 1–120.

Nagel, T. (1986), *The View From Nowhere* (New York: Oxford University Press).

Nunn, J.A., *et al.* (2002), 'Functional magnetic resonance imaging of synesthesia: activation of V4/V8 by spoken words', *Nature Neuroscience*, **5** (4), pp. 571–5.

Ramachandran, V.S. and Hubbard, E.M. (2001), 'Synaesthesia: a window into perception, thought, and language', *Journal of Consciousness Studies*, **8** (12), pp. 3–34.

Siegel, R.K. and Jarvik, M.E. (1975), 'Drug induced hallucinations in animals and man', in *Hallucinations: Behavior, Experience and Theory*, ed. R.K. Siegel and L.J. West (New York: Wiley), pp. 81–62.

Smilek, D. and Dixon, M.J. (2002), 'Towards a synergistic understanding of synaesthesia: combining current experimental findings with synaesthetes' subjective descriptions', *Psyche*, **8** (01), [http://psyche.cs.monash.edu.au/v8/psyche-8-01-smilek.html]

Uttal, W.R. (2001), *The New Phrenology: The Limits of Localizing Cognitive Processes in the Brain* (Cambridge, MA: MIT Press).

Varela, F.J. (1996), 'Neurophenomenology', *Journal of Consciousness Studies*, **3** (4), pp. 330–49.

Varela, F.J. and Shear, J. (ed. 1999), *The View From Within: First-Person Approaches to the Study of Consciousness* (Exeter: Imprint Academic).

Victor, M., Ropper, A.H. and Adams, R.D. (2000), *Principles of Neurology*, 7th edn. (New York: McGraw Hill), or see any standard neurology textbook.

Zaidel, E., Iacoboni, M., Berman, S. and Bogen, J.E. (2003), 'The callosal syndromes', *Clinical Neuropsychology*, 4th edn., ed. K.M. Heilman and E. Valenstein (New York: Oxford University Press), ch. 14.

Anthony J. Marcel

Introspective Report
Trust, Self-Knowledge and Science

This paper addresses whether we have transparent accurate access to our own conscious experience. It first sketches the origin and social history of this issue in the seventeenth century, when the trust one can have in self- knowledge was disputed in the religious, social and scientific domains. It then reviews evidence (a) that our conscious experience is disunified in several ways and has two levels, can be opaque to us, and contains much that is non-explicit; and (b) that attending to one's experience not only affects and changes it, but may also bring about specific content and phenomenal experience. Quite apart from (mis)trusting other people's introspective reports, we cannot trust ourselves even in knowing our own consciousness. Finally, several ways of coping with the problems produced by these characteristics are suggested.

Introduction

The recent research and theoretical literature demonstrates the legitimation of the content of conscious experience as topic and data and why it is important to psychology; and these two special journal issues presuppose these things. The arguments I have previously made for these (Marcel, 1988) do not need to be rehearsed here. Various questions concerning phenomenal reports and introspective evidence are addressed by others in these present journal issues. I shall concentrate here mainly, but not exclusively, on one question — the apparently absurd question of to what extent one can be accurately aware of the content of one's consciousness. The question is not just whether the scientist or philosopher can get a firm grasp on specific experience or on other people's experience, but whether one can get a firm grasp on one's own experience. Whether or not we can trust others, can we trust ourselves? The question seems absurd because it is often assumed, indeed it is almost definitional, that the one thing we know perfectly is the content of our own consciousness. The role of such experience as data for cognitive science I leave largely to others here, although I will comment on it. However, as will become clear, the question of self-knowledge is itself part of cognitive science. Difficulties in introspection and self-knowledge tell us about the nature and working of the mind, about kinds of knowing. Indeed one of my themes here is different kinds of knowledge and awareness. Concerning the relation between the concepts in the title of this paper, my argument is not only that we cannot unquestioningly trust our introspection, but that there are intrinsically subjective non-observational forms of self-knowledge whose contents are hard to articulate even to ourselves, and that although these are typically rejected by natural science (*naturwissenschaft*) a science of consciousness cannot ignore them.

Our principal access to another's consciousness is their introspective report, in whatever way it is made (Marcel, 1993). In what follows I shall enumerate several ways in which one does not have direct or transparent access to *one's own* conscious experience and the reasons for this. I shall end by suggesting some ways in which investigators of consciousness can cope with this problem. One of these relies on the proposition that in certain states one does have direct access to one's experience. However, first I will sketch extremely briefly some of the social historical background to the question of the directness and reliability of our acquaintance with our own consciousness.[1]

The intellectual positions held by scientists are embedded in longstanding attitudes and traditions which suffuse and are reflected in many social domains. Contextualising our current debates helps us to a richer understanding of our positions. The topic of this paper and of others here, the trust that can be put in certain kinds of knowledge, is by no means arcane and has a long and illuminating historical context. Although the issue long predates the renaissance, during the years of the Reformation and Counter-Reformation the peoples of Europe fought bitterly over it, shedding much blood. Indeed in certain societies the issue is still active, especially wherever one group seeks to control not merely the behaviour of the population, but its thought and self-confidence. The same issue was central to both religion and early modern science. What I shall call the '*paradox of protestantism*' was well recognised by the early seventeenth century; it lay in the individualism begot by renaissance humanism and the Reformation. The paradox was as follows. It was increasingly held that the surest knowledge was that gained by an individual through his or her own senses and experience. Such

---

[1] What I want to do here is to hint at ('only to hint at' because of space limitation, but also 'importantly to hint at') the cultural sources and pervasiveness of the ideas that I touch on in this essay. In this sense these following couple of paragraphs on history are allusions — allusions that are meant to open a door to further thought in the reader. The ideas and issues in this essay are not merely scientific and philosophical issues, nor even only tacit paradigm assumptions, but are concerns that lurk in the dark in us and that affect us, just as they acted as stage managers to social arrangements and the violent religious wars of the sixteenth and seventeenth centuries. Some scientists want to avoid such violence by restricting themselves to empirical, 'objective' and rational approaches to specific decidable topics, a form of positivism. (Donald Broadbent explicitly owned to this view.) I respectfully submit that an alternative with more hope of success is to do that, but also to confront and acknowledge longstanding attitudes to such topics that manifest themselves in social history. In this context Mary Midgley's recent book *The Myths We Live By* (2003) is most relevant. Perhaps this footnote is a central point of my essay. Certainly this essay is doing more than one thing. I make no apology for that.

knowledge has two characteristics. It is empirical as opposed to knowledge handed down by authority or the mediation of someone with special access, e.g. a priest. But it is also essentially knowledge *by acquaintance* (*connaissance*) rather than knowledge *by description* (*savoir*), and is therefore essentially private and subjective rather than public and objective. Late mediaeval and renaissance neo-platonism believed in the interdependence of finding truths about the natural world and the moral and spiritual state of the enquirer (Yates, 1964; 1979). For alchemists such as John Dee (and Isaac Newton) the quest was for all three as a unity: natural knowledge, spiritual knowledge and one's moral state go hand in hand and mutually guarantee each other. The gnosticism of the renaissance magi, alchemists and early protestant sects[2] was reflected in the principles of early modern science and the Royal Society. 'Truth' lay in trust — or rather with those individuals one could trust — which is why science could only be conducted by 'gentlemen' (Shapin, 1994). It was not just that credibility in science lay in the gentlemanly codes of trust, honour and integrity. It was also that the 'integrity' of the gentleman, his wholeness and internal one-ness, the transparent unity of his consciousness, what Schaffer in discussing these issues (1992; 1998) calls 'the continuity of self', guaranteed his probity, moral soundness and, most relevantly, accurate self-knowledge. He could trust himself, know accurately his own experience and knowledge (whereas others, e.g. women or servants, could not). But the paradox I referred to is that if such private knowledge was to be non-hermetic and widely shared it required a descriptive form that made it public and a way to match one person's experience with another's. And this is one of the problems confronted by John Locke in 1691 (Locke, 1691/1959).

Locke envisaged that the private nature of our experience cannot be publicly guaranteed. This is the problem of the inverted spectrum. It might be the case that the physical conditions that lead in me to the experience (not the brain state) that I call red lead in you to the experience that you also call red but which I call (and know as) green; and there is no way to decide. (Locke himself was not only concerned here with privacy or communication and how words refer, but rather more

---

[2] There is an interesting contrast and similarity in the gnosticism of the neoplatonic alchemical magi and the protestant sects. While the former sought direct apprehension of divine knowledge through crystal ball gazing, trance states and visions, they treated texts hermeneutically, as requiring interpretation, following the traditions of the Kabbala. Some kinds of knowledge are not manifest and require the adept to acquire the keys. (The Kabbalist tradition of treating man as a text to be read is reflected in the symbolic interpretation of psychoanalysis. The politics of psychoanalytic therapy depend on whether or not the analyst offers authoritative interpretation to which he or she has privileged access.) By contrast the protestant sects sought a fundamentalist approach to the Bible as the word of God. If it requires no interpretation, then anyone who can read has direct access to divine knowledge. Some sects continue to prioritise individual divine inspiration over textual knowledge and in an egalitarian organization with no priest (consider Quaker traditions). We can see here how disputes over the degree of cognitive mediation required in different kinds of knowledge are continuous with arguments over justifications for universal or restricted democratic franchise, arguments vividly and passionately made in the Putney debates of 1647–8 (Woodhouse, 1938) between the parliamentary army, radicalised by their protestantism, and their commanders involved in government, principally Cromwell. Current political debates in Western societies over trust and access to knowledge in various domains echo the seventeenth century. The early currency of these issues in England can be seen in Shakespeare's *The Merchant of Venice* and *The Tempest* and Ben Jonson's *The Alchemist*.

with the nature of the content of thought and sensation. In fact he concluded that the problem has no great practical importance. However the upshot and its relevance to the present issue is unaffected.) Locke found no way to untangle this problem, and Daniel Dennett has recently (1991) taken the problem to be insoluble and used it to justify his instrumentalism and his belief that all one can do, and all there is, is 'heterophenomenology'. (Dennett's point that phenomenal experience is not independent of cognition and attitude is echoed in this paper; but the idea that it is not 'raw' does not lead me to his conclusion that, because something is not what it was thought to be, it does not exist.) If Locke's position is valid, then observations that depend on or involve personal phenomenal experience cannot be compared between individuals. The answer developed by scientists is to depend on 'objective' measuring devices (which merely shifts the location of the Cartesian boundary between subjective and objective). I have argued (Marcel, 1988) that there are methods that can go some way to compare individuals' experiences (at least for experiences that exist within a domain, such as colour or taste), namely what is called second-order isomorphism. Essentially one carries out psychophysical studies where people are asked to make relational judgements that focus on their subjective experience of pairs of stimuli from a given domain such that one can derive a reliable multidimensional map of the stimulus-provoked experiences in that domain for each individual and then compare individuals. In this way one might at least examine whether the relative tastes on various dimensions (bitterness, creaminess) of a range of chocolates, i.e. a map of the tastes, are the same for two or more individuals, even if one cannot characterise exactly how each one tastes to each individual.

However, when I wrote that I was assuming what Locke assumed, that at least our private experience, the content of our experiential or conscious states (as opposed to what is not conscious), is transparent to each individual mind. This assumption is held by many today — I know what the soup in my mouth tastes like now; I know what my present visual experience is; I know what I am thinking at this moment, I know when I am in pain. However, this assumption has been challenged in a number of ways (see Dennett, 1991); and I myself no longer hold the assumption, for both empirical and conceptual reasons. Some of these reasons amount to rejecting the assumption of what Schaffer (1992; 1998) calls the continuity of self: that our consciousness is unified. I shall try to make these reasons and concepts clear, as well as try to give some answer to the challenge. In fact, as indicated above, even in the seventeenth century the problem was recognised, since only certain kinds of people could be trusted, not only to report faithfully, but to have clear access to their own states. Psychologists of the late nineteenth century Wurzburg school believed that it was worth having subjects who were *trained* in introspection. (Contemporary psychophysicists also use subjects who are trained, but to *avoid* relying on their consciousness in order to avoid contamination of supposedly mind-independent sensations!) Indeed today we think that individuals differ in how well they know their own states of mind, especially when they are in certain emotional states. (Ironically, it is often the clear-headed, detached individuals who are least acquainted with their own mental states.)

The problem is a serious one. If we think that phenomenal experience is of any relevance to psychology or cognitive science or plays any role in our mental life or behaviour, then we will have to have some grasp on it. I have argued that it plays several causal roles, in self-monitoring, learning, task formation, emotional control, and voluntary action (Marcel, 1988). To take the last example, Dickinson and Balleine (2000), Jack and Shallice (2001) and I (Marcel, 1988; Lambie and Marcel, 2002) have argued that intentional actions of certain types rely on certain kinds of conscious content, specifically phenomenal content.[3] If we are to test such hypotheses properly we need to know the content of conscious states of people who perform the relevant actions or not or form the relevant intentions or not. How well can we or our experimental subjects gain such knowledge?

Self-Trust and Self-Deception: Can We Get a Firm Grasp On Our Own Experience?

Many psychologists indeed suppose that when they require people to report their experience or base intentional responses on it, if the subjects are not lying then the content of their awareness is faithfully reflected by their report or action. Others do acknowledge that there are problems, but mostly only when it concerns vague or uncertain phenomenology. But quite apart from practical problems in reporting our conscious experience or obtaining objective characterisations or measures of it, there are two kinds of problem in knowing our own experience. One of these concerns the non-unitariness of consciousness: consciousness appears to be non-unitary in several ways; the other concerns the effects on experience of attending to it. The first problem questions whether we can trust ourselves. The second problem questions whether phenomenal experience and the content of consciousness are the kind of things on which a firm grasp can be had.

1. *Disunities, opacity and inexplicitness*

If our consciousness were single and unitary it would be transparent to us. However, much suggests that it is non-unitary, and in several ways.

(i) *Different Levels*. First, when most psychologists distinguish between conscious and nonconscious states or between consciousness of something and its lack, they conflate two distinctions in the single dichotomy. There is a logical distinction between two aspects of what is referred to as consciousness, (a) experiential phenomenology or phenomenal experience (what it's like) as opposed to nonphenomenal states, and (b) awareness of something (a kind of knowing — by acquaintance) as opposed to its lack. This does not *per se* imply any distinction

---

[3] My proposal was that the reason for the necessity of phenomenal status for components of intentions for action (e.g. the object on which an action is to be performed) lies in our normative rationality, for justification of the action (to intend to grasp that cup if 'that cup' is not phenomenally present to me would be meaningless), and is therefore culturally contingent. However, there may be a more interesting reason. I have endorsed (Marcel, 2003) the view that we have a non-observational awareness of our voluntary actions. Insofar as we do, it may be *constitutive* of intentions for action that they have phenomenal status.

between levels; the former may just be an aspect of the content of the latter. However, as John Lambie and I have argued (Lambie & Marcel, 2002), several other considerations do suggest such a separation between levels or aspects of consciousness. If you are asked to report the sensations in a body part to which you were not attending and of which you were not already aware, were there any such sensations before attending to it? When a pain attracts attention to itself and its location, is it the painfulness, the hedonic phenomenology, that attracted attention or is it some subpersonal information? Either the act of attending brought about the phenomenology or the phenomenology preexisted attention and awareness of it. While attention itself and the way one attends affects the phenomenological content of awareness (see below), several phenomena make a case for separating first-order nonreflective consciousness (of x) and second-order reflective awareness (of x), both being distinct from nonconscious representation (of x). This distinction has long been held in the phenomenological tradition, and has variously been made recently in analytic approaches by among others Nelkin (1989), Farthing (1992), Marcel (1993), Block (1995), Lane (2000), and in a similar spirit by Schooler (2002). (However, meta- cognition and reflection, focussed on by Schooler, are different from what I refer to as awareness, and might be considered third-order.)

Several kinds of empirical phenomena imply the reality of this distinction. Some patients with hemiplegia after stroke who are apparently unaware of their paralysis (anosognosia for hemiplegia) appear to experience the proprioceptive phenomenology of movement and its lack in the affected limbs, but also are simultaneously aware and unaware of such occurrent experience when questioned appropriately (Marcel *et al.*, forthcoming). Some patients when asked 'Is this arm weak? Do you have any problem with movement?' reply 'No', but when asked 'Is this arm ever naughty? Does it ever not do what you want?' eagerly assent. Some patients when they try to lift their plegic arm on request declare immediately after that they are unable to do so, but if asked at the same time a generic question, 'Can you move both arms normally?' or 'Is this arm weak?', deny any problem. Some patients are unaware of their plegia on all conversational criteria but never attempt bilateral actions, whereas others show the opposite, complaining about their plegia yet frequently having accidents in attempting bipedal or bimanual actions. Whether these phenomena are due to different forms of knowledge or to dissociative consciousness, they reflect dual awareness. And the fact that patients are unaware of such contradictions suggests co-occurrent split awarenesses. This split awareness of sensation also shows itself in other cases. In the 'hidden observer' phenomenon in hypnosis (Hilgard, 1977), if people with hypnotically induced unawareness of some class of sensation (audition, pain) are questioned appropriately they report awareness of the sensation (they are both aware and unaware of it), but *as in another person* ('She has a pain.' 'He can hear you.'). In general anaesthesia with centrally acting analgesics, patients often later recollect, or even re-experience, a pain they were apparently unaware of during the operation. Further research suggests that under general anaesthesia centrally anaesthetised patients are frequently simultaneously aware and unaware of sensations as they occur, depending on appropriate methods of questioning (see Marcel, 1993). Characterisation of such phenomena entails a distinction between phenomenal experience and awareness of it. This distinction is also illustrated by and explains the separate occurrence of blindsight (loss of conscious visual experience with preserved nonconscious vision), Anton's syndrome (unawareness of blindness), and Anton's syndrome with bilateral blindsight (Marcel, 1995). Blindsight patients have normal access to their visual experience, which in their case is absent in the scotomic area, which is why they deny seeing; yet they are shown to have nonconscious vision by accuracy of guessing and by the reliable effects of stimuli in the blind field on action and on other stimuli (Weiskrantz, 1990; Marcel, 1998). In Anton's syndrome unawareness of blindness is plausibly due to lack of access to absent visual experience. In the third case, a patient assessed as totally bilaterally blind after traumatic impaction of the occipital poles was unaware of his blindness for 7 months. Yet during this time his ability to point to light targets and guess luminance levels correctly and other indirect tests all indicated blindsight, i.e. nonconscious vision without visual phenomenology. Apparently he lacked access to the latter's absence. The difference between blindsight and unawareness of blindness appears to be in one's awareness of visual phenomenology or its loss. The presence of these three conditions and normal vision suggests a dissociation between nonconscious vision, visual phenomenology, and awareness of the latter.

Recently John Lambie and I (Lambie & Marcel, 2002) have reviewed evidence that conscious emotion experience can consist of the phenomenological aspect of emotion states alone or of the additional second-order awareness of such experience. We also argued that in strategies coping with anxiety, in defence mechanisms such as repression, in anger disorders, and in ordinary cases of world-focussed emotion, there is strong evidence of the presence of emotion phenomenology without, or with deficient, second-order awareness of it. Present spatial constraints prevent a review of the evidence and arguments here. However we argue that it is only by the separate existence of two levels of consciousness that repression (and other defence mechanisms) is possible. I have suggested (Marcel, 1988; Lambie & Marcel, 2002), as have Dickinson and Balleine (2000) and Jack and Shallice (2001), that certain intentional actions are dependent on phenomenal representation of the object of the intention. Specifically, Lambie and I proposed that the representation of a self-state, event or object has to have phenomenal status if it is to be the basis for intentional action regarding it (e.g. avoiding something that provokes anxiety), though second-order awareness of the representation is not required. But without such awareness individuals will be unable to give a veridical account of their action. Initial testing by Lambie & Baker (forthcoming) of our prediction for 'repressors' has proved successful. Such individuals are defined on the basis of conventional criteria (low self-reported trait anxiety but high defensiveness; Weinberger, 1990). It was found that they can base appropriate intentional action on anxiety about specific events (e.g. by avoiding the event), anxiety they are unaware of, but they are unable to account appropriately for their action. This finding was predicted by our two-level proposal, but runs counter to Jack and Shallice's (2001) hypothesis that intentional action requires prior awareness of that which becomes the content of intention. The only cases that they consider as nominal exceptions are where awareness is disunified 'horizontally' in dissociative states; see (ii) below. Our two-level proposal would save their general position regarding repression. The conclusion that phenomenal representation of something is necessary for intentional action involving it can also be drawn from the fact that people with blindsight do not make spontaneous intentional actions with regard to stimuli falling in their blind field though there is good evidence for the nonconscious representation of such stimuli. This also reinforces that what is deficient in blindsight is visual phenomenology rather than visual awareness.

The same distinctions seem to be demanded by psychophysical data I have reported for normal people (Marcel, 1993), some of whose essential relevant features have been replicated by Overgaard and Sørenson (submitted). In the relevant conditions subjects either (a) reported whether or not they had a conscious experience of detecting a luminance increment (in a specified location) or (b) rapidly guessed the presence or absence of such a luminance change (irrespective of their conscious experience of it); and in both cases they did this using three response modes simultaneously: eye-blink, manual button-press and oral 'yes/no'. The relevant finding was that in the condition requiring report of experience, but not in the guessing condition, the responses on any one trial frequently dissociated across the different response modes. That is, often subjects were saying 'Yes, I am aware of a luminance increase' with their eyes, but were saying *at the same time* 'No, I am not aware of a luminance increase' with their manual gesture. Report requires awareness, whereas guessing does not; and the observed greater accuracy of guessing was consistent with that. Although subjects correctly remembered the instructions, they did not realise, even when questioned immediately after a trial, that there was a discrepancy between their reports in the different modes. If the report condition reflects conscious representation while the guessing condition does not, as is widely believed, then the disjunction between response modes in report suggests simultaneous awareness and unawareness of the visual change. And if this is to be explained by differential access to such representation from that which underlies report, then this seems to imply two levels of consciousness, with the second-order level underlying report and being equivalent to the kind of knowing that is awareness. (The alternative to splits in a second-order awareness is to postulate multiple sensations/representations of a single event, one for each arbitrary response mode – which is unprincipled to say the least. It is also possible that it was the *intention* to report which affected the perceptual representation, and did so prior to introspective access. But since there is no plausible reason why it should do so differentially for the different response modes, this interpretation is weakened. See Marcel, 1993, for fuller discussion of procedure and theoretical inferences.)

Although the phenomena briefly sketched above may each have alternative explanations, taken together they make a strong case for the distinction between two levels of consciousness, apart from what is nonconscious: first-order phenomenal experience and second-order awareness. In our proposal (Lambie and Marcel, 2002) the former can exist independently of the latter, which can take the former as its content. As such, first-order experience is articulated in terms of Gestalt laws of organization, but may be otherwise non-coherent. Second-order awareness is underlain by focal attention, and is what supports reflection, report and later purposive recollection of episodic memory; it tends to logical and conceptual coherence. We proposed empirical criteria to distinguish the three levels or statuses. (1) What one is aware of is *reportable*; (2) what is phenomenological but without awareness is *expressible*, in behaviour and manner; (3) what is neither of these but is nonconscious is *indexible by indirect means* (e.g. by its influence in priming or biasing of some further nominally unrelated task which avoids asking the subject directly about what one is interested in). Regarding (2), it is a widely held tenet that 'expression' reflects phenomenal experience, and this can exist without awareness of such experience. A good example is where someone is shouting aggressively but is genuinely unaware of their anger, which is frequent, especially in anger disorders; there is certainly something it is like to be in such an emotional attitudinal state.

Clearly, the relation of second-order awareness to first-order phenomenal experience has an affinity with Rosenthal's (1993) notion of higher-order thoughts. However, for us this mechanism does not account for consciousness as a whole but only for awareness. It is hard to see how higher order thoughts about X bestow phenomenal status on X. The main alternative position to our trichotomy is that there is a single distinction between conscious and nonconscious, and that what people cannot report is outside of experiential consciousness. Our arguments against this are (a) empirical dissociation between our three criteria, (b) the reality of the conceptual distinction between phenomenal and nonphenomenal states, (c) that a trichotomy gives a more satisfying account of a wide range of phenomena including those mentioned above.

(ii) *Splits within a Level of Consciousness*. Several of the phenomena mentioned above imply not just a distinction between two levels of consciousness, but also 'horizontal' splits within consciousness. In anosognosia for hemiplegia, the hidden observer in hypnosis, and in the equivalent in general anaesthesia, people seem to be simultaneously aware and unaware of an aspect of their phenomenal experience. However this comes about, it implies differential access to that experience. Likewise, in the psychophysical experiments reported by myself (1993) and by Overgaard and Sørenson (submitted) differences in report modalities show up in conditions requiring report or, as Overgaard and Sørenson term it, introspection. These cases imply differential access to an experience. But differential access in turn implies separation or splits within that from which there is access, ie. that which is the basis of report: second-order awareness. The inference is that second-order awareness can be non-unitary or split. In fact in dissociative states such as fugue, post-traumatic stress disorder and multiple personality disorder this is indeed what is claimed. If awareness is split, one awareness would not have access to the content of another (at least not symmetrical access). If it did there would be no dissociation or incoherence in reported content. It follows that anosognosic patients, normal subjects in the psychophysical experiments with different report modes, and multiple personalities 'could appreciate that there is a contradiction in their behaviour (if pointed out to them) but could not experience the contradiction directly' (Marcel, 1993, p 179). Indeed the assumption of unity of consciousness is difficult to shake off precisely because logically we can never directly experience more than one consciousness at a time. Like split-brain patients, we could only infer them from external observation.[4]

(iii) *Non-unification within a Single Consciousness.* Even if dissociated awarenesses coexist in a mind, it is usually supposed that at least the contents of a single consciousness are smoothly unified over space and time, i.e. that there is a referential wholism to experience. Certainly it *seems* that there is; we rarely notice disjunctions in it. However, consideration of certain phenomena undermine even this supposition. Regarding space, Duncan (1984) briefly displayed two overlapping objects each varying on two dimensions; subjects had to report the values of two features, either both on one object or one on each. People were simultaneously aware of both features and their relation on a single perceptual object but were not simultaneously aware of two features belonging to separate objects or their relation, even when there was no spatial distance between them. The equivalent point is made by our perception of impossible objects such as the Penrose triangle or depictions by Escher. We are not aware at a single moment of the incoherence of their features; we have to focus on each part in turn and calculate their (in)coherence as 3-D objects or scenes. Again, in the visual motion aftereffect (the waterfall illusion) in the test stimulus we see the middle part of the field moving (in the opposite direction to the previous inducing stimulus in that part of the field), but at the same time we see the whole, including the critical part, as not moving; yet we do not usually notice the incoherence of our experience. Regarding time, auditory streaming, as in polyphonic music or simultaneous speech streams, provides good examples. A sequence of tones can be presented at a rate where, if they are heard as a single stream, the temporal order of successive tones is clearly and accurately perceived. But if at the same rate the pitch range of alternate tones is separated such that the sequence is heard as two independent streams, one has no determinate temporal experience of the order of successive tones between the two streams (Bregman and Campbell, 1971). Experientially the two streams are not temporally unified. Thus it appears that neither the spatial nor temporal relations between perceptual objects is determinately unified in experience. It would seem that this is the norm. All perceptual experience entails segmentation into figure and ground, but no determinate relation between the two is experienced and they do not cohere. To that extent, all perceptual consciousness is dis-unified. It is often asserted that focal attention creates

---

[4] My present approach is to take many of the phenomena suggesting split awareness at face value and infer corresponding divisions in consciousness. That is, I have implicitly reified the phenomena. However, many of them (though not the psychophysical experiments nor the distinctions between blindsight, Anton's syndrome and normal vision) may be a matter of the person's construal of their strange state, or a coping strategy, or even just a manner of speaking.

unity in consciousness, eg. by binding or integrating. However, even if it creates unity within its focus, it creates disunity in perceptual consciousness as a whole. Moreover, we do not usually notice such noncoherence.

(iv) *Opacity, Recessiveness, Non-observational Awareness, and Pragmatics*. In addition to the various apparent problems of disunities, there is another set of ways in which our conscious experience is not internally transparent, or not equally so.

(a) Much of our experience is perceptually recessive or in the background in so far as it is articulated into figure and ground. If we are asked to report our experience, we are aware of and report a small part of it. If, as we suggested earlier, what we are aware of is a function of selective focal attention, much of our experience is simply unattended. At any moment our phenomenal experience is multimodal and (differentially) complex in any one modality. We simply do not notice most of it unless it is aberrant. For example, what it's like to be sitting in this chair as I write has a particularity, but I would only notice it if it were somehow different from normal. A single episode of emotion often consists in bodily sensations, thoughts, evaluative external perceptions and action urges. Yet when people are asked to report their experience they tend to be aware of only subsets of these (Lambie, submitted). As Gallagher (1986) pointed out, in external perception the experience of our bodies is present but usually as a spatial background to awareness of the world. What we do not attend to, what is not figural, is experientially relatively unarticulated in awareness.

(b) Many features of our phenomenology, far from being explicit, are implicit (in the true sense of the term) such that we do not notice them. This has nothing to do with selective attention. Neither the perspectivalness of perceptual experience nor our perceived egocentric self-location are explicit in such experience. Gibson (1979) pointed out that egoreception (perception of the self) and exteroception (perception of the world) go hand in hand. Perceptual information specifies a spatial entity that Neisser (1993) calls 'the ecological self', which consists not just in one's spatial viewpoint, but also one's relative place, movement and posture and one's spatial abilities (e.g. Can I reach that object with my hand?). Yet this self is perceptually recessive: it is not a segmented perceptual object; so much so that Hume (1739) famously failed to be perceptually aware of it. Ownership of sensations, percepts and other conscious content is an important feature of our experience, yet we are hardly ever explicitly aware of it and it is virtually never reported. It is only in special states, e.g. of detachment, or when pathology gives the experience of dis-ownership that we notice it, as in felt disownership of action in Anarchic Hand (Della Sala *et al.*, 1994; Marcel, 2003) or disownership of limbs in anosognosia for plegia (Gerstmann, 1942). Indeed there are many aspects of our phenomenology that we do not notice and that are not explicit in our awareness which form precisely what phenomenologists call 'nonreflective awareness' (Merleau-Ponty, 1962). To realise and investigate such properties or features of our phenomenology one cannot rely on report; one has to think about it. Ownership of action provides a good example. In a recent paper (Marcel, 2003) I sketched various forms that the content of awareness of our own actions might take. Introspection yields little insight to arbitrate on this. I proposed that a crucial aspect of normal felt ownership of action is non-observational awareness of action (i.e. where one does not perceive the action from outside it, but is aware of it as part of oneself, the experiencer and actor; O'Shaughnessy, 1980), and that felt *dis*ownership of action in Anarchic Hand syndrome is partly due to the patient being restricted to an observational awareness of their anarchic actions 'from outside the action' (see below). If one engaged in detached introspection of one's awareness of action this would precisely remove one from the non-observational awareness that characterises our normal phenomenology. One would search one's phenomenology in vain, like Hume looking for his 'self'.

It is unfortunate that current cognitive science has little recourse to the concept of non-observational knowledge, which in its first-person subjectivity captures an aspect of immersed phenomenology. Cognitive science's models of awareness are either cybernetic monitoring, higher-order thoughts about the content, access to a further representational stage, or activation above a threshold, all of which are third- personal and involve no subjectivity. This is partly connected with its lack of concern with phenomenology and differences in kinds of knowledge. The nearest it comes to these topics is in Tulving's original conception of episodic memory, in simulation theory of social understanding, and in the Gibsonian theory of direct perception, where there is no epistemic gap between knower and known. The importance of this is as follows. Intrinsically first-person forms of awareness are a problem for naturalistic approaches, including cognitive science, and are often dismissed as unscientific. But since such forms of awareness have a role in science and in testing theories about the nature of such awareness, dismissing them as subjective or mere 'folk psychology' to be displaced by greater rigour is not an option. That is, science not only needs to take subjectivity as a topic, it needs somehow to incorporate subjectivity as a method.

(c) One also has to beware that certain features of our phenomenal experience are *pragmatically* backgrounded. A good example of this is to be found in emotion experience. Almost everyone agrees that hedonic tone (pleasant/unpleasant) is a pervasive (and motivating) feature of emotion experience. Yet if people are asked to report their experience during emotions, either at the time or retrospectively, very few respondents spontaneously mention it, though they will do so if later probed specifically. Lambie and Marcel (2002) suggested that 'it seems that sometimes hedonic tone is taken to be too obvious to report or to be an entailed property of the emotion reported or that the assumed task demand is to report only substantive or discrete objects of experience' (p. 226). However, it may be that pragmatics may affect not just report but salience in experience itself. The implication of this is that taking reports of experience is an interactive endeavour, a point that Jack and Roepstorff (2002) have made; but, further, that an individual's experience or the explicitness of its content is socially dependent (a notion inherent in psychoanalytic procedure for helping latent content become manifest to the experiencer herself).

2. Attention affects and creates experiential content

All the aspects of our consciousness sketched above make it very difficult to know our own experience. But there are yet other kinds of problem. These lie in the very act of attending to our own experience, of gaining second-order awareness of it, whether to examine it, remember it or report it.

(i) *Attention can influence its object.* Attending to one's experience, introspecting, changes the content, nature and form of the experience. It is also widely accepted that the content, nature and form of the experience that constitutes the content of awareness depends on the way that we attend. John Lambie and I (2002) have recently emphasised what we call the *mode* of attention, the manner in which one attends at any time — an aspect of attention stressed by William James (1890) but largely ignored by most current psychology. It is a constitutive feature of focal attention and encompasses two related but distinct dimensions. Analyticism–syntheticism refers to the extent to which attention is directed to components or to a whole, to a lower- or higher-level description or category, or to significance. (It is related to the idea of local versus global attention but is less tied to spatiality.) The more analytic one's attention, the more one's experience is abstracted and decontextualised, consisting of separate components. One may attend to speech as a series of phonemes or in terms of words or meaning. One may attend to one's emotion in terms of separable bodily sensations and the situational specifics or at the other extreme in terms of an emotion category such as simply 'anger'. Immersion–detachment refers to one's stance toward the object of attention. It is the attentional counterpart of the mood of verbs in language — the difference between indicative, conditional, subjunctive. If one is totally engrossed in an activity or in one's perceptual world one has no phenomenal separation from it. It has phenomenal truth: it just 'is'. Examples of detachment are reflective observation of oneself or one's activity, and observation of the world as something separate, unaffecting and unaffected by oneself; this is often called being 'cool'. Such an attitude underlies propositional attitudes and doubt. By altering their attentional mode depressed patients in cognitive behaviour therapy may come to apprehend their feeling of worthlessness as merely an unrealistic thought, but their immediate sense of worthlessness is nonetheless compelling. Differences in the mode of attention yield different phenomenology. On both its dimensions pathology can render individuals restricted to one or other extreme, and the content of their awareness differs accordingly (see Lambie and Marcel, 2002, for review). Whether there is a first-order phenomenology prior to the effects of the mode of attention or whether all phenomenal experience is subject to it is theoretically moot .

(ii) *Attention can create its object.* Even more problematically, attending to one's experience can create or add to the experience. Consider the case of imagery. If one has a mental image and is then asked a question about the content of that image, one may assume that the content of the experience which one then reports pre-existed the mental enquiry one made about the image. If one is asked to visually imagine a woman's face and is then asked about her earrings or the colour of her lipstick, one may give a definite answer. But usually until the question is asked the image contains no information as to earrings or lipstick: an image contains only what the imager has imagined and only what is canonically necessary. An enquiry one makes of one's image itself creates the parameter enquired of and supplies a value. As Dennett (1991) has pointed out, there is no sure way in which we can know whether that information was in the image prior to the enquiry or not (though indirect techniques may be informative).

(iii) *Awareness distorts its object*. Lambie and Marcel suggested two aspects of the content of consciousness that create further difficulties for apprehending our own experience. We characterise first-order phenomenology as it presents itself (though not theoretically) as nonanalytic and nondecomposable, and therefore apparently ineffable, inducing people to capture it by analogy and metaphor. It can also be non-coherent (as illustrated above by the lack of smooth unification). By contrast we characterise second-order awareness as subject to one's conceptual structure, tending to coherence and logicality. For these reasons, awareness, especially when analytic and detached, tends to distort at least some of our first-order phenomenology. The allusive language of imaginative writers (Edgar Allen Poe at one extreme, Virginia Woolf at another more analytic extreme) or the nondeterminate spatiality depicted by painters like Delacroix or Howard Hodgkin resonate with and capture our first-order experience in a way that detached academic language does not.

(iv) *Our theories can mask our experience*. A further problem is that our tacit theories about our experience may intervene in our apprehending it. And it is often difficult to distinguish our theories of what it should be from the experience itself. This has been a difficulty for those involved in cross-cultural research, e.g. on emotion experience. People report feeling 'hot' when angry, etc., according to their folk theories and language. Another example is that until Nigro and Neisser's (1983) paper, many people never considered the possibility that we may see ourselves as a visual object in recollected episodic memories; indeed many theorists considered it impossible. The consequence of such a theoretical assumption was that nobody spontaneously reported seeing themselves in remembered episodes. But when people were explicitly asked to report whether they saw themselves in memory images, many people did so, prompting Nigro and Neisser to coin the terms 'field memories' and 'observer memories'. The prima facie problem is to avoid mistaking our theory of experience for the experience itself. The problem behind that problem is whether there always is such a distinction.

The phenomena and considerations outlined above (for further evidence and argument, see Lambie & Marcel, 2002) suggest that far from being singular, unified and transparent, consciousness is disunified in several ways, much of its content is opaque, and it is affected by, even the product of, paying attention to it. (However I have also argued that in certain states and situations, e.g. immersed states, our consciousness is unified; Gallagher and Marcel, 1999) The problems implied by these characterisations undermine the notion of sure and transparent acquaintance with one's own conscious experience.[5] Even if not definitive, the suggestions above provide a reason why we cannot have *unquestioning* trust in what introspection yields. The one thing that Descartes was convinced he could trust was his acquaintance with his own consciousness. We should trust ourselves even less. Descartes' Method of Radical Doubt was not radical enough.

Coping Strategies

'Coping strategies' is the expression used by clinical psychologists and those concerned with emotion to describe the ways we deal with what provokes anxiety, is unpleasant, or hard to tolerate, including both defence mechanisms and therapies. I use the term here to refer to how to deal in research with the problems created by what is sketched above for capturing the content of our consciousness. However, if one takes seriously the radical lack of a basis for self trust, one might find oneself in a state of real existential anxiety. This may be the reason why the most commonly used strategy has been a denial of the problem. However, there are alternatives.

Researchers can try to minimise the performance constraints of report in various ways. They can induce subjects to deal with only one aspect of their conscious experience at a time. They can require subjects to make simple responses such as button-presses on automatic recording devices. The question then arises as to when such responses are functionally equivalent to communicative speech acts of report or not (e.g. button press = 'The two stimuli look equally bright now.') I have discussed this issue elsewhere (Marcel, 1993). Researchers can try to manipulate subjects' attitudes by speeded or non-speeded responding or by asking for guesses about external stimuli versus reports of experience. They can use a range of psychophysical procedures. They can compare reports made in various ways. However, the problems outlined above concern what would seem

---

[5] Jack and Roepstorff (2002) draw a distinction between reports of what one perceives and reports of one's experience, calling them first- and second-order. They say that the former cannot be mistaken (are incorrigible), but the latter can be. While I sympathise with the distinction, I disagree with both conceptions. First, as I interpret Jack and Roepstorff, they conflate world-focussed experience versus self-focussed experience with first-order versus second-order, a conflation also made by Sartre (1939/1962) and other phenomenologists. The two distinctions are separate. One can have second-order awareness of a world-focussed percept, as when one attends to and contemplates the colour of a flower with the online (nonverbal) reflective thought 'How blue it is!' Second, evidence reviewed by Lambie and Marcel (2002) suggests that neither kind of content is transparent to second-order awareness nor immune to error or distortion in awareness. If Jack and Roepstorff mean that first-order experience is incorrigible by definition, there is no counterargument. But if they do not mean that, then there is room for doubt. One can be mistaken about one's experience just as one's experience can be mistaken. In Dennett's (1991) discussion of practised coffee tasters, it may be impossible to distinguish whether the current taste is different and less liked than the former taste or whether the original taste is the same but now unpleasant. Although I suggested a method of distinguishing the two options (Marcel, 1988), the case nonetheless undermines the incorrigibility of first-order experience. However, I agree with Tony Jack (personal communication) that when one's mode of knowing is non-observational ('being in the experience' rather than representing it) our knowledge of our experience is direct and cannot be mistaken. I hope I have not misrepresented Jack and Roepstorff here. Our terminology is different and may be a source of confusion; and we largely agree.

to be prior to report, people's access to their own conscious experience and the dependence of such experience on knowing it. As such they provide major challenges to anyone who attempts to capture the content of our experience or awareness, and they bedevil almost all approaches, experimental, clinical or otherwise. I believe that there are alternative strategies and that one might use them as converging operations. It is worth pointing out that although neuroscience data can obviously be useful (as I shall mention below), neural indices *per se* of conscious content rely on prior correlation with phenomenal reports that are subject to the problems above. Moreover, activation in brain imaging of a structure usually associated with a certain experience is no guarantee of the experience, and to equate neural vehicle with mental content requires a very strong and as yet unwarranted assumption (though one that is pragmatic).

The first strategy is simply to use more than a single behavioural criterion. This is what was done in our research on anosognosia for hemiplegia (above), where different kinds of question and behaviour reflected a non-singular awareness or unawareness of the patients' paralysis.

The second coping strategy is to attempt to infer the content and form of a certain kind of experience from converging use of empirical data of various kinds and conceptual analysis. This is what I have done in a recent paper on the sense of agency (Marcel, 2003), in order to specify whether the phenomenal content itself of immersed concurrent awareness of one's action contains the ownership of the action or not, and if so, in what way. (Since I was concerned with voluntary action rather than movement or involuntary reflex, ownership refers to who is the agent.) An alternative to this possibility is that felt ownership (self-agency) is inferred or is a product of subpersonal cybernetic comparison (e.g. between feedback and intended outcome or motor specifications). Phenomenal reports give no purchase on this issue, largely because what constitutes felt ownership of action is not explicit in the phenomenology to which one has access, and because paying attention to it alters the normal immersed condition under which one is aware of one's actions. My method was to integrate different approaches: normal behavioural data on awareness of intention and movement (mainly from vibrotactile illusions of limb position and movement); data on the functional and neural locus of initial awareness of action; the phenomenology of brain stimulation and of several clinical conditions whose functional neuroanatomy is reasonably known, especially felt disownership of actions in Anarchic Hand syndrome; as well as conceptual insights from philosophy and theoretical psychology. This enabled combining several insights: awareness of action often does not derive from the action itself; initial awareness of action appears to be underlain by specifications for bodily movement following intention but preceding motor commands, and the Supplementary Motor Area is involved in this; in disownership of action in Anarchic Hand the common lesion site is the Supplementary Motor Area; specifications for bodily action are necessarily in an egocentric frame of reference that unifies points of origin of action on the body, which are the motor equivalents of the perspectival point of origin in perception (the 'ecological self'). In brief, the ensuing proposal was that experience of ownership of actions underlies our occurrent immersed sense of agency, and it is given in the phenomenal content of awareness of action itself. This content consists in the specifications for movement, and the owner is given as their spatially perspectival source. That such specifications follow intention but precede motor commands explains why awareness of actions can often be inaccurate. Pathological experience of disownership of action is not due to cybernetic failure but to awareness of action being restricted to being observational (outside of the action); this can be due to damage to structures providing non-observational phenomenology of emerging action or to the attentional stance one takes. The same investigative strategy of converging operations can be applied to other instances of phenomenal content.

The third set of strategies is aimed at the problem of accurately apprehending first-order phenomenology. Normally we can only know it via second-order awareness, which usually transforms it. However there are states, eg. of immersion, where second-order awareness is minimal and where the content of consciousness approximates first-order phenomenology. It is possible to reinstate such states by episodic recollection. Therefore the possibility exists of either catching one's phenomenal experience 'out of the corner of one's eye' when in such states or via memory. Hurlburt and Heavey (forthcoming) offer an alternative strategy for catching one's immersed phenomenology 'out of the corner of one's eye'. They advocate random (thus unexpected) time sampling of occurrent experience by use of a bleeper to prompt subjects to report their immediate experience. I endorse the usefulness of this technique, because subjects do not pre-prepare themselves to report on each occasion and are less likely to be in a detached state at the time. But I also agree with Hurlburt and Heavey's caution that the method will often fail to evade the problems I have explicated in this paper. As another alternative, there are states called 'flow experiences' (Csikszentmihalyi, 1978) when people are highly focussed and immersed in an activity in which they have attained a certain level of skill. In such states we can be simultaneously both highly immersed and also detachedly aware of our immersed phenomenology. Meditation techniques used in therapy (Teasdale, 1999) can sometimes be used to attain such states. Training in such techniques offers a possibility of circumventing normal limitations. When people are trained in 'mindfulness' techniques they appear to become more receptively, rather than effortfully, aware of both internal and external experience and less transforming of it. Such states appear to approximate pre-reflective and non-reflexive consciousness. Two good reviews of the conceptual basis and practical applications of such training are provided by Baer (2003) and Brown and Ryan (2003). To quote the latter, 'rather than generating accounts about the self, mindfulness "offers a bare display of what is taking place" (Shear and Jevning, 1999, p. 204).' One has to be cautious lest one is merely inducing a different state of mind rather than more accurate access; but the fact that in such states people seem aware of more than otherwise does suggest a closer approximation to one's first-order phenomenology.

A fourth approach is to attempt to specify the content of experience and the nature of subjectivity on theoretical and logical grounds, much as in the continental tradition of technical phenomenology and that of analytic philosophy, but allied to empirical science. Examples of this are provided by the work of Shaun Gallagher (Gallagher and Meltzoff, 1996), Vittorio Gallese (2001), Josef Parnas and Dan Zahavi (see Zahavi and Parnas, this issue) and Naomi Eilan (1995). One reason for recourse to such approaches is that insights and concepts are provided by these traditions that are simply not forthcoming from the purely naturalistic approach of information processing and cognitive science. Although the different languages and conceptual frameworks of these approaches may currently amount to a cultural barrier and seem incommensurable, intercourse if not integration is possible, and what I am suggesting is supplementation. Indeed converging operations need not be restricted to methods within a discipline but can be interdisciplinary. However, it should be noted that while such alternative approaches can help to specify contents of experience and awareness in general, they do not help the scientist gather data on particular experiences of specific content at particular times. For this, the first three strategies or equivalent alternatives are necessary.

My proposals clearly go beyond the methods that most cognitive scientists or neuroscientists feel comfortable with. Yet I see no essential problem with them nor any discontinuity with cognitive science. But if one is to go beyond the most minimalist studies of consciousness, if one is to confront the nature of specific mental and phenomenal content, one has to have recourse to a range of appropriate methods. And here is one of the problems with subjectivity and objectivity. What 'objective science' too often means to its practitioners is not just a matter of methodology; it is a matter of assuming that the object of study has a thing-like mind-independent reality. We cannot assume that this is the case when the object of study is phenomenal experience and the content of consciousness. If we assume that first-order consciousness is something independent of subjectivity and that the only problem arises when we try to become aware of it, try to introspect, then we are assuming the Cartesian Theatre, as Dennett (1991) has called it. Certainly some people in the seventeenth century, as my introduction indicates, believed that one's phenomenal experience is something to which one can have 'accurate' access and which therefore has objective existence. If we are to emerge from the seventeenth century, or perhaps accept our continuity with it, we have to accept the intrinsic subjectivity and mind-dependence of phenomenal experience and the content of consciousness.

Acknowledgements

I would like to thank Simon Schaffer for discussion of seventeenth century science and politics, and Stephen Butterfill for comments on philosophical and theoretical issues. I am also grateful to Andrew Lawrence, Hugo Spiers, Peggy Postma and Fiona Clague for reading and commenting on an earlier version of this manuscript. The editors and reviewers have helped to clarify and sharpen the paper. All remaining faults are mine.

References

Baer, R.A. (2003), 'Mindfulness training as a clinical intervention: A conceptual and empirical review', *Clinical Psychology: Science and Practice*, **10**, pp. 125–43.

Block, N. (1995), 'On a confusion about a function of consciousness', *Behavioral and Brain Sciences*, **18**, pp. 227–47.

Bregman, A.S. and Campbell, J. (1971), 'Primary auditory stream segregation and perception of order in rapid sequences of tones', *Journal of Experimental Psychology*, **89**, pp. 244–9.

Brown, K.W. and Ryan, R.M. (2003), 'The benefits of being present: Mindfulness and its role in psychological well-being', *Journal of Personality and Social Psychology*, **84**, pp. 822–48.

Csikszentmihalyi, M. (1978), 'Attention and the holistic approach to behaviour', in *The Stream of Consciousness*, ed. K.S. Pope & J.L. Singer (New York: Plenum Press).

Della Sala, S., Marchetti, C. and Spinnler, H. (1994), 'The anarchic hand: A fronto-mesial sign', in *Handbook of Neuropsychology, vol 9*, ed. F. Boller and J. Grafman (Amsterdam: Elsevier, North Holland).

Dennett, D.C. (1991), *Consciousness Explained* (London: Little, Brown and Co.).

Dickinson, A. and Balleine, B.W. (2000), 'Causal cognition and goal-directed action', in *The Evolution of Cognition*, ed. C.H. Heyes and L. Huber (Cambridge MA: MIT Press).

Duncan, J. (1984), 'Selective attention and the organization of visual information', *Journal of Experimental Psychology: General*, **113**, 501–17.

Eilan, N. (1995), 'Consciousness and the self', in *The Body and the Self*, ed. J. Bermúdez, A. Marcel, N. Eilan (Cambridge, MA: MIT Press).

Farthing, G.W. (1992), *The Psychology of Consciousness* (Englewood Cliffs, NJ: Prentice Hall).

Gallagher, S. (1986), 'Body image and body schema: A conceptual clarification', *Journal of Mind and Behavior*, **7**, pp. 541–54.

Gallagher, S. and Marcel, A.J. (1999), 'The self in contextualized action', *Journal of Consciousness Studies*, **6** (4), pp. 4–30.

Gallagher, S. and Meltzoff, A. (1996), 'The earliest sense of self and others: Merleau-Ponty and recent developmental studies', *Philosophical Psychology*, **9**, pp. 213–36.

Gallese, V. (2001), 'The "Shared Manifold" hypothesis: From mirror neurons to empathy', *Journal of Consciousness Studies*, **8** (5–7), pp. 33–50.

Gerstmann J. (1942), 'Problem of imperception of disease and of impaired body territories with organic lesion', *Archives of Neurology and Psychiatry* (Chicago), **48**, pp. 890–913.

Gibson J.J., (1979), *The Ecological Approach to Visual Perception* (Boston, MA: Houghton Mifflin).

Hilgard, E.R. (1977), *Divided Consciousness: Multiple Controls in Human Thought and Action* (New York: Wiley).

Hume, D. (1739), *A Treatise of Human Nature*, ed. L.A. Selby-Bigge (Oxford: Clarendon Press, 1888/1975).

Hurlburt, R.T. and Heavey, C.L. (forthcoming), 'To beep or not to beep: Obtaining accurate reports about awareness', in *Trusting the Subject II*, special issue of *Journal of Consciousness Studies*.

Jack, A.I. and Shallice, T. (2001), 'Introspective physicalism as an approach to the science of consciousness', *Cognition*, **79**, pp. 161–96.

Jack, A.I. and Roepstorff, A. (2002), 'Introspection and cognitive brain mapping: From stimulus-response to script-report', *Trends in Cognitive Sciences*, **6** (8), pp. 333–9.

James, W. (1890), *The Principles of Psychology* (1981 edition, Cambridge, MA: Harvard University Press).

Lambie, J.A. (submitted), 'The world and the self in emotion experience' Manuscript submitted for publication.

Lambie, J.A. and Baker, K.L. (forthcoming), 'Intentional avoidance and social understanding in repressors and nonrepressors: Two functions for emotion experience?', *Consciousness and Emotion*.

Lambie, J.A. & Marcel, A.J. (2002), 'Consciousness and emotion experience: A theoretical framework', *Psychological Review*, **109**, pp. 219–59.

Lane, R.D. (2000), 'Neural correlates of conscious emotional experience', in *The Cognitive Neuroscience of Emotion*, ed. R.D. Lane and L. Nadel (Oxford: Oxford University Press).

Locke, J. (1691/1959), *An Essay Concerning Human Understanding*, ed. A.C. Fraser (New York: Dover).

Marcel, A.J. (1988), 'Phenomenal experience and functionalism', in *Consciousness In Contemporary Science*, ed. A.J. Marcel and E. Bisiach (Oxford: Oxford University Press).

Marcel, A.J. (1993), 'Slippage in the unity of consciousness', in *Experimental and Theoretical Studies of Consciousness*, Ciba Foundation Symposium 174, ed. G.R. Bock and J. Marsh (Chichester: John Wiley).

Marcel, A.J. (1995), 'Anton's syndrome with blindsight', Paper presented at the 11th European Workshop on Cognitive Neuropsychology, Bressanone, Italy, January, 1995.

Marcel, A.J. (1998), 'Blindsight and shape perception: Deficit of visual consciousness or of visual function?', *Brain*, **121**, pp. 1565–88.

Marcel, A.J. (2003), 'The sense of agency: Awareness and ownership of action', in *Agency and self-awareness*, ed. J. Roessler and N. Eilan (Oxford: Oxford University Press).

Marcel, A.J., Tegnér, R. and Nimmo-Smith, I. (forthcoming), 'Anosognosia for plegia: Specificity, extension, partiality and disunity of bodily unawareness', *Cortex*.

Merleau-Ponty, M. (1962), *Phenomenology of Perception*, tr. Colin Smith (London: Routledge).

Midgley, M. (2003), *The Myths We Live By* (London: Routledge).

Neisser U. (1993), 'The self perceived', in *The Perceived Self: Ecological and Interpersonal Sources of Self Knowledge*, ed. U. Neisser (Cambridge: Cambridge University Press).

Nelkin, N. (1989), 'Unconscious sensations', *Philosophical Psychology*, **2**, pp. 129–41.

Nigro G. and Neisser, U. (1983), 'Point of view in personal memories', *Cognitive Psychology*, **15**, pp. 467–82.

O'Shaughnessy, B. (1980), *The Will*, Volume II (Cambridge: Cambridge University Press).

Overgaard, M. and Sørenson T.A. (submitted), 'Introspection as a process distinct from first order experience'.

Rosenthal, D.M. (1993), 'Thinking that one thinks', in *Consciousness*, ed. M. Davies & G.W. Humphries (Oxford: Blackwell).

Sartre, J.P. (1939/1962), *Sketch for a Theory of the Emotions*, tr. P. Mairet (London: Methuen).

Schaffer, S. (1992), 'Self evidence', *Critical Inquiry*, **18**, pp. 327–62.

Schaffer, S. (1998), 'Regeneration: The body of natural philosophers in Restoration England', in *Science Incarnate: Historical Embodiments of Natural Knowledge*, ed. C. Lawrence and S. Shapin (Chicago: Chicago University Press).

Schooler, J.W. (2002), 'Re-representing consciousness: Dissociations between experience and meta-consciousness', *Trends in Cognitive Science*, **6**, pp. 339–44.

Shapin, S. (1994), *A Social History of Truth: Civility and Science in Seventeenth Century England* (London: Chicago University Press).

Shear, J. and Jevning, R. (1999), 'Pure consciousness: Scientific exploration of meditation techniques', in *The View From Within*, ed. F. Varela and J. Shear (Exeter: Imprint Academic).

Teasdale, J.D. (1999), 'Emotional processing, three modes of mind, and the prevention of relapse in depression', *Behaviour Research and Therapy*, **37**, pp. S53–S57.

Weinberger, D.A. (1990), 'The construct validity of the repressive coping style', in *Repression and Dissociation*, ed. J.L. Singer (Chicago: University of Chicago Press).

Weiskrantz, L. (1990), 'Outlooks for blindsight: Explicit methodologies for implicit processes', The Ferrier Lecture, 1989 [Review], *Proceedings of the Royal Society of London. B. Biological Science*, **239**, pp. 247–78.

Woodhouse, A.S. (1938), *Puritanism and Liberty* (London: Dent).

Yates, F.A. (1964), *Giordano Bruno and the Hermetic Tradition* (London: Routledge & Kegan Paul).

Yates, F.A. (1979), *The Occult Philosophy in the Elizabethan Age* (London: Routledge & Kegan Paul).

Zahavi, D. and Parnas, J. (2003), 'Conceptual issues in the study of infantile autism: Why cognitive science needs phenomenology', *Journal of Consciousness Studies*, **10** (9–10), pp. 53–71.